AF664492

A Practical Manual

FUNDAMENTALS OF HORTICULTURE

THE AUTHOR

Dr. Suneeta Singh is presently serving as Associate Professor and Head at the Department of Horticulture, School of Agricultural Sciences, Shri Guru Ram Rai University, Dehradun, Uttarakhand. She is associated with teaching and research in the courses pertaining to horticulture for the last 12 years. She did her B. Sc. (Agriculture) from Institute of Agriculture Sciences, Banaras Hindu University, Varanasi, U.P. in 2005. She did her M.Sc. (Ag.) Horticulture in 2007 and earned her Ph.D. in Horticulture from G. B. Pant University of Agriculture & Technology, Pantnagar, Uttarakhand. The author was recipient of ICAR-JRF Fellowship in M.Sc. She also qualified ICAR-NET thrice. Dr Singh started her professional career in 2010 after earning her Ph.D. degree as Senior Research Fellow in Department of Vegetable Science, G.B.P.U.A. &T., Pantnagar. Dr Singh was also worked as Research Associate in Vivekananda Parvatiya Krishi Anusandhan Sansthan (VPKAS), ICAR, Almora, Uttarakhand. She has published 2 Books, 24 Book chapters, 45 Popular articles covering various aspects of Horticultural technologies in various magazines and more than 52 Research papers in peer reviewed UGC-CARE listed as well as NAAS rated scientific journals of International and National reputes. She has supervised and mentored 32 M.Sc. (Ag.) Horticulture students. Under her guidance 6 Ph.D. students are enrolled. In recognition of her services to the scientific journal, the author was selected as Chief Editor, editorial board member, review board member and reviewer of several National and International journals. She is a peer guy dedicated to the cause of development to horticulture.

A Practical Manual

FUNDAMENTALS OF HORTICULTURE

Dr. Suneeta Singh

M.Sc.(Ag.), Ph.D., ICAR-NET
Associate Professor & HOD
Department of Horticulture
School of Agricultural Sciences
Shri Guru Ram Rai University
Dehradun-248 001, Uttarakhand
(India)

2024

Daya Publishing House®
A Division of
Astral International Pvt. Ltd.
New Delhi – 110 002

© 2024 AUTHOR

ISBN: 978-93-5461-948-9

Publisher's Note:

Every possible effort has been made to ensure that the information contained in this book is accurate at the time of going to press, and the publisher and author cannot accept responsibility for any errors or omissions, however caused. No responsibility for loss or damage occasioned to any person acting, or refraining from action, as a result of the material in this publication can be accepted by the editor, the publisher or the author. The Publisher is not associated with any product or vendor mentioned in the book. The contents of this work are intended to further general scientific research, understanding and discussion only. Readers should consult with a specialist where appropriate.

Every effort has been made to trace the owners of copyright material used in this book, if any. The author and the publisher will be grateful for any omission brought to their notice for acknowledgement in the future editions of the book.

All Rights reserved under International Copyright Conventions. No part of this publication may be reproduced, stored in a retrieval system, or transmitted in any form or by any means, electronic, mechanical, photocopying, recording or otherwise without the prior written consent of the publisher and the copyright owner.

Published by : **Daya Publishing House®**
A Division of
Astral International Pvt. Ltd.
– ISO 9001:2015 Certified Company –
4736/23, Ansari Road, Darya Ganj
New Delhi-110 002
Ph. 011-43549197, 23278134
E-mail: info@astralint.com
Website: www.astralint.com

ACKNOWLEDGEMENT

Ever since the introduction of new course for professional B.Sc. (Hons.) Agriculture degree, there was a dire need to have a practical manual on Horticulture subject which covers the practical syllabus of undergraduate B.Sc. (Hons.) Agriculture. The present manual designed according to the syllabus offered in undergraduate degree of Agriculture as per fifth Dean Committee for the student which covers the practical with objectives, material required, procedure and steps to follow precautions to be taken, observations to be recorded and exercise to be done by the students. The main objective of this manual is to acquaint students about the brief knowledge of practicals. We hope that users will find the manual immensely useful.

We look forward to receiving the valuable suggestions of readers for improvement of this manual.

Dr. Suneeta Singh
(Associate Professor & Head, Horticulture)

PREFACE

This practical guide would be very helpful for students because it includes all the necessary material in-depth and frees up more time for practical exercises. Because they allow students to combine scientific concepts and aesthetics of art to a few horticultural tasks, such potting a plant, horticultural practical's are frequently of considerable interest. The pupils' practical expertise will help them in their future employment. The "A Practical Manual - Fundamentals of Horticulture" contains a full description of the field activities. The exercises in this handbook were created using the updated recommended syllabus. It is envisaged that the Practical Manual on Fundamentals of Horticulture will be an excellent teaching tool for undergraduate students. The information presented in this manual is considered to be of utmost value to the students. The manual, which consists of 21 chapters, is an in-depth and thorough compilation of all the facts and numbers related contemporary Fundamentals of Horticulture. However, humankind is aware of phenomena outside of the ecosphere as well. There is a potential that there will be gaps and shortages. Thus, there is always room for improvement. I will be very grateful to every reader who extends their goodwill in helping to improve the current endeavour.

I owe a debt of gratitude to my professors at the Department of Horticulture at the Institute of Agricultural Sciences, BHU, Varanasi, as well as the Department of Horticulture at the G.B. Pant University of Agriculture and Technology, Pant Nagar, who helped us begin the shaky discussion of the topic and whose blessings allowed us to complete the manuscript. I lack the words to adequately describe how grateful we are to our devoted parents, whose encouragement and blessings forced us to finish this endeavour. I want to publicly express my gratitude to my son Mr. Avyaan and my spouse Dr. Anil for supporting us and keeping us free to work on our literature. I want to thank everyone whose actions have been beneficial, whether directly or indirectly.

(Suneeta Singh)

CONTENTS

Acknowledgement *v*

Preface *vii*

1. Study About Horticulture Garden Tools and Implements 1
2. Identification of Horticultural Crops 8
3. Preparation of Nursery Bed and Raising Seedling 15
4. Raising of Seedling in Portrays 21
5. To Study About Growing Media 24
6. Study of Pots, Potting, Depotting and Repotting 28
7. Study of Orchard Layout 31
8. Study About Orchard Planting Systems 33
9. Propagation Through Specialized Vegetative Structures 40
10. Propagation by Seed in Fruit Plants 43
11. Study on Propagation by Cutting 45
12. Study on Propagation by Layering 49
13. Study on Propagation by Grafting 52
14. Study on Propagation by Budding 57
15. Study About Training and Pruning 60
16. Study of Manure and Fertilizer Application 68
17. Preparation of Growth Regulators Substance 75
18. To Study about Herbaceous, Shrubbery Border and Plant Components of Garden 78
19. Lawn Management 88

20.	**Irrigation Management**	**93**
21.	**Study of Different Types of Greenhouses Based on Shape**	**96**
22.	**Study on Hydroponics**	**100**
23.	**Study on Vertical Gardening**	**106**
	Terminology	**114**

Exercise No. 1

OBJECTIVE - Study About Horticulture Garden Tools and Implements

Understanding the numerous garden implements and tools used for various horticultural operations is crucial. To attain maximum efficiency and a speedy completion of work, a person should be knowledgeable about the use of the appropriate sort of tool or implement for a given task, along with proper maintenance, repairs, and storage of tools and implements. Some tools are easy to use and are used for straightforward tasks. However, some operations need for specialized equipment. Therefore, it is crucial to exercise/perform various horticultural operations from the stage of land preparation through harvesting with the appropriate selection of acceptable tools/implements.

The following list includes some of the instruments, tools, and plant protection gear needed for various horticultural operations:

S. No.	Name	Uses
1.	Axe	Used for felling trees and cutting branches and pruning
2.	Bill hook	Commonly used for cutting the big branches/stems near the ground surface or to remove the old and dead branches
3.	Budding knife	Used for budding purpose
4.	Budding-cum-grafting knife	It has two sharp blades for budding and grafting specially with the back end made up of brass used to lift or loosen the bark for inserting the bud
5.	Pick Axe	Used for digging hard, compact and stony soil, loosening the soil, pit digging, opening of trenches
6.	Cultivator	It is a tractor drawn implement used for tilling the soil

S. No.	Name	Uses
7.	Carpenter's saw	To prune the thick and bigger branches and useful in crown grafting
8.	Crow-bar	An iron rod usually of 1.5m in length and 2.5 to 4.0 cm thick with one end pointed and the other end flattened. Used for digging pits in hard soil, breaking stone and moving rocks
9.	Chain saw	A petrol and electric operated machine use for cutting thick branches sharply.
10.	Digging fork	Used for loosening the moist soil and mixing manures in pits
11.	Drainage hoe	Used for making the drainage channel and to remove silt deposit in the channels.
12.	Dibbler	To make small holes on the seed beds in order to place seeds or transplant seedlings
13.	Duster	Hand operated machine use for dusting of chemicals which are in the form of powder and machine has capacity of 5, 10, 15 kg generally.
14.	Forester's shear	To prune the medium sized branches (4-8cm) which are at higher height on the trees
15.	Garden hand rake	For collecting stump and other residues of plant, removing stubbles, small stones, leveling of nursery beds and breaking clods, formation of small beds
16.	Garden trowel	For lifting more number of seedlings
17.	Garden weeder	Used for removal of weeds from garden
18.	Grass shear	To cut the outgrowth of grasses planted in posts, carpet beds
19.	Hand fork	Used for hoeing, compost handling and to break the clods.
20.	Hand cultivator	To loosen the soil, remove clods, pebbles in nursery beds and mixing of manures and fertilizers.
21.	Hand leveller	Used in small bed and nursery for levelling land and covering the seed after sowing.
22.	Hatchet	To remove or cut down the bigger stems and broken stems
23.	Hedge shear	To prune the tender parts of garden shear the plants, it is especially useful for trimming hedges, borders, topiary work
24.	Hose pipe	To irrigate flower beds, lawns etc.
25.	Kurpi	For weeding and stirring the soil in the pots and beds.
26.	Looper	Used for looping of large plants. This tool has long handles with extra leverage for cutting hard thick stems.

S. No.	Name	Uses
27.	Lawn mower	To cut the grass uniformly in the lawn. It is having a roller behind to pad the grass to have cushion.
28.	Lawn sprinkler	For irrigating lawns
29.	Pruning saw	Used to prune the thicker branches (4-6cm girth) of an acute crotch (angle)
30.	Pruning knife	For pruning of thicker branches and it has curved knife.
31.	Pruning shear	For cutting small sized branches.
32.	Plant support (stakes)	Bamboo and wooden sticks of different sizes are used to provide support to the plants (staking) after planting during their initial stages of growth.
33.	Polythene strip	Plastic strips having different thickness are used as tying material in plant propagation.
34.	Pot	Pot of different shapes and sizes are used in the nursery. Clay, asbestos, cement, plastic and metallic pots of different shapes and sizes are usually used.
35.	Pro tray (Seedling tray)	Seedling tray is used for raising of seedlings and hardening of micro propagated plants. The size of the tray varies with the number of cavities ranging from 25-250.
36.	Rotary weeder	For cutting of grasses in lawn, carpet beds, edges etc.
37.	Secateur	Used to prune the branches, twigs, water suckers etc. of small plants.
38.	Sickle	For cutting grasses, vegetables etc.
39.	Scythe (Dabba)	For cutting lawn grasses, vegetables etc.
40.	Spade	To loosen the soil, prepare irrigation channels, collect the soil in heaps and facilitate filling up of soil, manure etc. in the baskets.
41.	Soil sieve	Tool use for straining the soil, FYM, vermicompost etc. for removing of pebbles, plant debris and other dead plant parts from media mixture.
42.	Transplanting trowel	To lift the young seedlings along with a boll of earth for transplanting.
43.	Tree pruner	To cut down the smaller branches of the trees without climbing.
44.	Trenching hoe	For light collection of soil, irrigation purposes and opening of trenches.
45.	Tree calipers	To measure the girth of trees trunks.
46.	Tasla or Pan	A multipurpose pan use for lifting plants in nursery, measure the soil, FYM and sand for preparation of mixture of media.
47.	Water can with rose head	To water the young seedling in seed beds, potted plants etc. the rose head facilitates with fine spray of water which prevents the washing down of soil.

S. No.	Name	Uses
48.	Weed cutters	Special kinds of weed cutter have a serrated double edged steel blade and handle about 60 cm long. It is used with swinging strokes in two directions.
49.	Wheel barrow	To transport manures, soil, seedlings, garden waste etc.
50.	Back pack sprayer	For spraying of soluble fertilizers / plant protection chemicals on nursery beds, flower beds, potted plants etc.
51.	Knapsack sprayer	For spraying nursery beds, flower beds, potted plants etc.
52.	Foot Sprayer	For spraying of soluble fertilizers/plant protection chemicals on nursery beds, flowerbeds, potted plants etc. It requires two labourers for operation.

Precautions

- Choose the right tool/implement for specific job.
- Check the tool/implement about its working condition before use.
- Clean them, if necessary, wash with water and dry it and oil them periodically.
- Store in the right way in the right place after use every time.

Some Garden Tools

		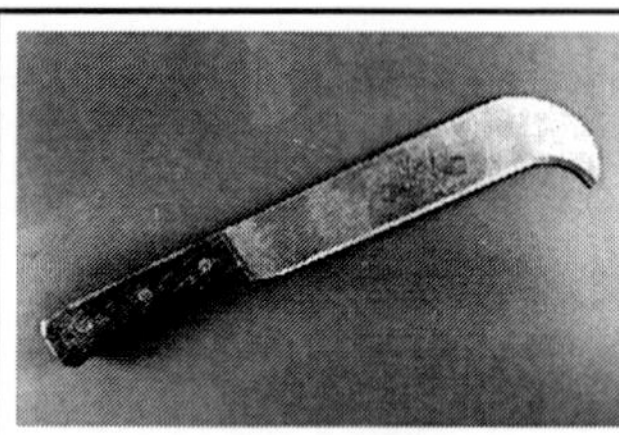
Axe	**Budding knife**	**Bill Hook**

Budding cum Grafting knife	**Pick axe**	**Carpenter's saw**

	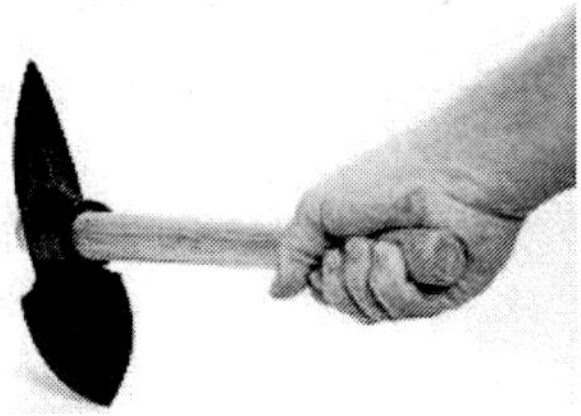	
Digging fork	**Hand Hoes**	**Drainage hoe**
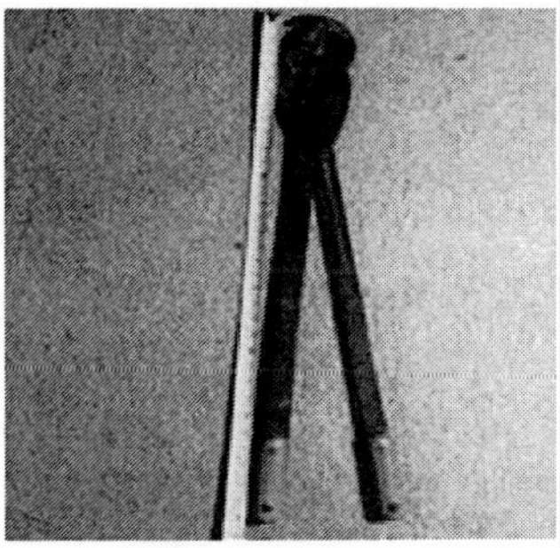		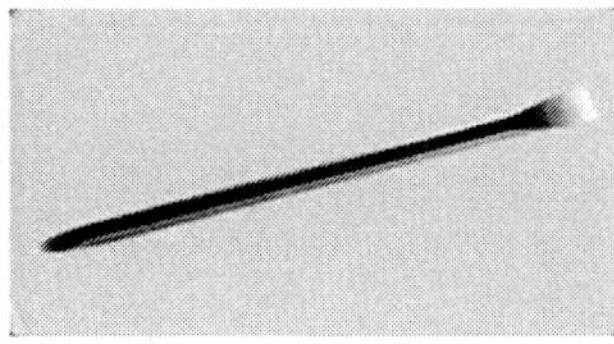
Forester' shear	**Cultivator**	**Crow-bar**
	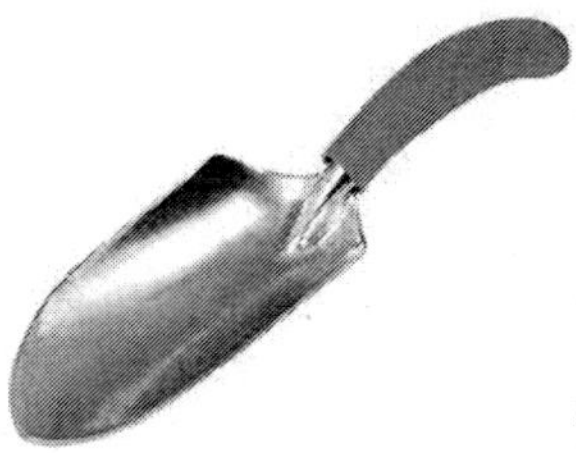	
Garden hand rake	**Garden trowel**	**Grass shear**
	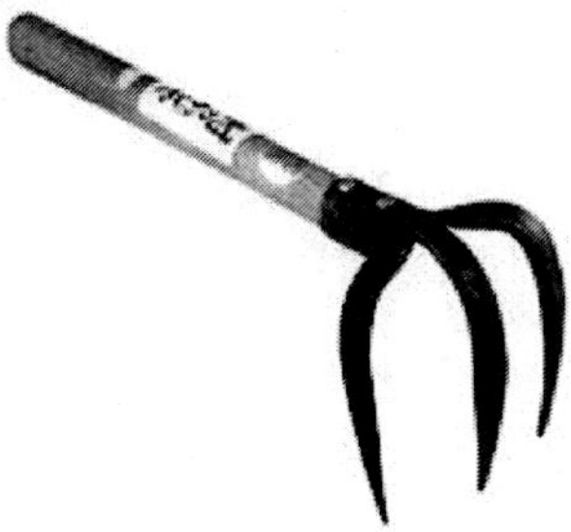	
Hand fork	**Hand cultivator**	**Hand laveller**

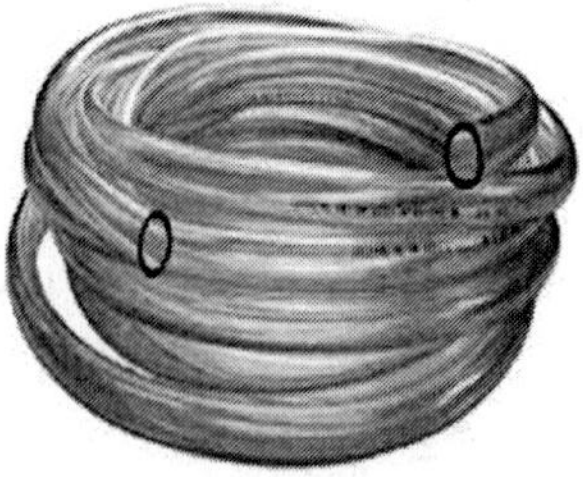	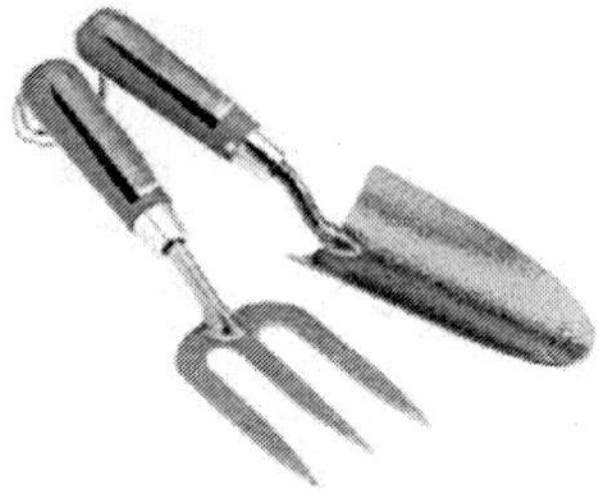	
Hose pipe	**Kurpi-varvari**	**Lawn mower**
	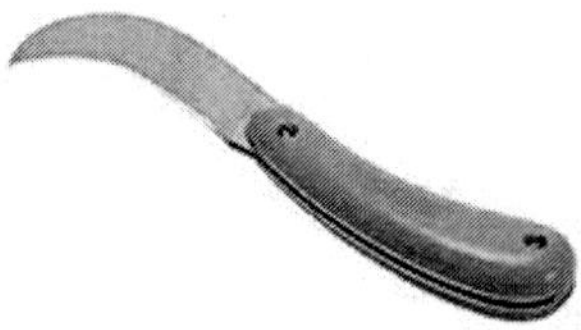	
Lawn Sprinkler	**Pruning knife**	**Pruning shear**
		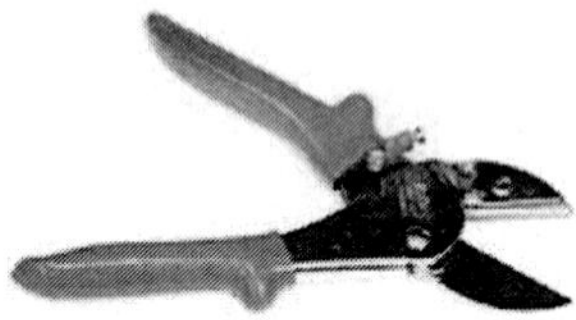
Pruning Saw	**Rotary weeder**	**Secateur**
	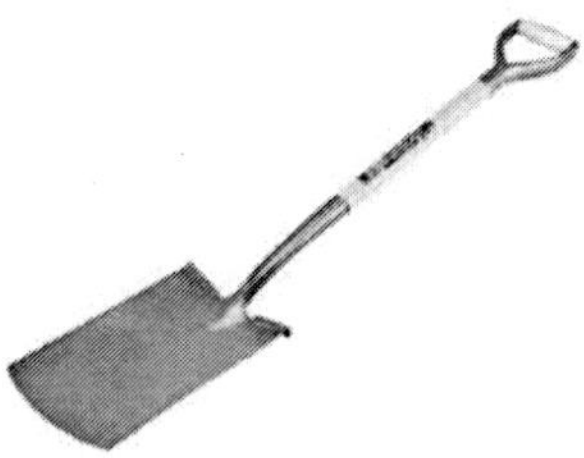	
Sickle	**Spade**	**Tree pruner**

Tree Calipers	**Rose can**
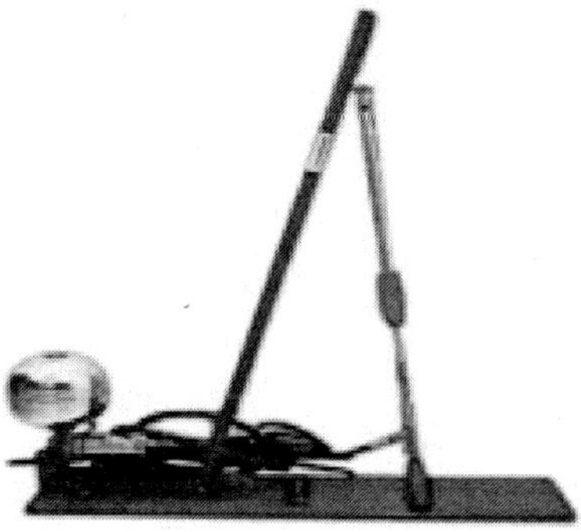	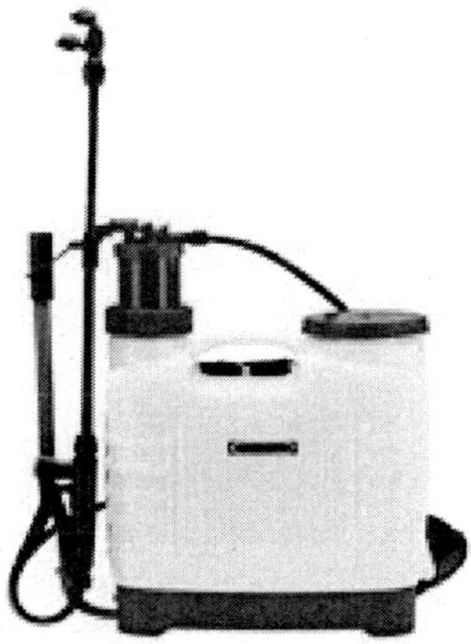
Foot sprayer	**Backpack sprayer**

Exercise No. 2

OBJECTIVE - Identification of Horticultural Crops

Horticultural crops are classified in different times for various purposes where each of the classifications have definite objective and grouped under botanical, horticultural and commercial heads.

BRANCHES OF HORTICULTURE

- **Pomology**- production and marketing of fruit crops.
- **Olericulure**- production and marketing of vegetable crops.
- **Floriculture and landscaping**- production and marketing of flower/ornamental crops, beautification through design and alteration of land using planting material etc.
- **Plantation Crops**- cultivation and marketing of commercial crops on extensive scale like coconut, tea, coffee, Cashew nut etc.
- **Spices & Aromatic Crops** - cultivation and marketing of crops having aroma, taste and flavor value.
- **Medicinal and aromatic Plants**- cultivation and marketing of crops having potential source of drugs.
- **Roots and Tuber crops**- cultivation and marketing of root crops like radish, carrot, turnip, sugar beet and tuber crops like potato etc.
- **Fruit & Vegetable Preservation**: - applied branch for protection and processing of horticulture produce to increase shelf life as well as preparation of value added products.
- **Plant propagation & nursery management**:- branch which deals with the production of multiple plants either sexual, asexual or vegetative plant parts means.

Nomenclature is a method or scheme for naming plants. Binominal is a universal method in which a plant is assigned a two-word name, one suggesting the genus or group it belongs to and the other separating it from other plants in the group. The plant is classified and named usingthis two-word or binominal appellation.

NOMENCLATURE OF FRUIT CROPS

S. No.	English Name	Common Name	Botanical Name	Family
1	Aonla	Amla	*Phyllanthus officinalis* Gaerth	Euphorbiaceae
2.	Almond	Badam	*Prunus dulcis* Mill	Rosaceae
3.	Apple	Seb	*Malus pumila* Mill	Rosaceae
4.	Apple Crab apple)	Chhota seb,	*Malus baccata* Borgen	Rosaceae
5.	Apricot	Khurmani	*Prunus armeniaca* L.	Rosaceae
6.	Avocado	Avocado	*Persea americana* Mill.	Lauraceae
7.	Bael	Bil	*Aegle marmelos* Correa.	Rutaceae
8.	Banana (edible fruit)	Kela	*Musa sapientum* L.	Musaceae
9.	Banana (cooking type)	Kela	*Musa paradisiaca* L	Musaceae
10.	Black berry	Brambles	*Rubus canadensis* L.	Rosaceae
11.	Bardados cherry	Bardados cherry	*Malpighia glabra* L.	Malpighiaceae
12.	Carambola	Kamrakh	*Averrhoa carambola* L.	Oxalidaceae
13.	Cashew nut	Kaju	*Anacardium occidentale* L.	Anacardiaceae
14.	Chinese jujube	Ber	*Zizyphus jujube* Lamk.	Rhamnaceae
15.	Cherry (Sour)	Sour Cherry	*Prunus cerasus* L	Rosaceae
16.	Cherry (Sweet)	Cherry (Sweet)	Prunus avium L.	Rosaceae
17.	Chestnut	Chestnut	Castanea sativa Mill.	Fagaceae
18.	Coconut	Khopa or Nariyal	Cocos nucifera L.	Palmaceae
19	Custard apple	Sita phal	Annona squamosa L	Annonaceae
20.	Date	Khajoor	Phoenix dactylifera L.	Palmacea
21.	Date (Wild)	Khajoor	Phoenix sylvestris Roxb.	Palmacea
22.	Fig	Anjeer	Ficus carica L.	Moraceae
23.	Gajanimma	Gajanimma	Citrus pennivesiculata Tanaka.	Rutaceae
24.	Gooseberry	Kiwi fruit	Actinidia deliciosa	Dilleniaceae
25.	Grape (American)	Angoor	Vitis lbrusca Bailey.	Vitaceae
26.	Grape (European)	Angoor	Vitis vinifera L.	Vitaceae
27.	Guava	Amrood	Psidium guajava L.	Myrtaceae
28.	Hazel nut	Bhatia badam	Corylus avellana Mill.	Betulaceae
29.	Indian jujube	Ber	Zizyphus mauritiana Lamk.	Rhamnaceae

S. No.	English Name	Common Name	Botanical Name	Family
30.	Jackfruit	Katehal	Artocarpus heterophyllus L.	Moraceae
31.	Jambolan	Jamun	Syzygium cuminii Skeels	Myrtaceae
32.	Karonda	Karonda	Carissa carandas L. A	Apocynaceae
33.	Lemon	Baramasi lemon	Citrus limon Burm.	Rutaceae
34.	Lemon	Galgal	Citrus limon Burm.	Rutaceae
35.	Litchi	Litchi	Litchi chinensis Sonn.	Sapindaceae
36.	Mandarin	Sangtra	Citrus reticulata Blanco.	Rutaceae
37.	Mandarin (Satsuma)	Sangtra	Citrus unshiu Marc.	Rutaceae
38.	Loquat	Loquat	Eriobotrya japonica Lindl.	Rosaceae
39.	Mangosteen	Mangosteen	Garcinia mangostana L.	Guttiferae
40.	Quince	Beedana	Cydonia oblonga Mill.	Rosaceae
41.	Peach	Aru	Prunus persica L.	Rosaceae
42.	Persimmon	Japani phal	Diospyros kaki L.	Ebenaceae
43.	Pomegranate	Anar	Punica granatum L.	Punicaceae
44.	Pecan nut	Pecan nut	Carya illinoinensis Koch	Juglandaceae
45.	Pistachio nut	Pista	Pistacia vera L.	Anacardiaceae
46.	Pear	Nashpati	Pyrus communis L.	Rosaceae
47.	Plum	Alubokhara	Prunus bokhariensis Schneid.	Rosaceae
48.	Strawberry	Strawberry	Fragaria chiloensis Duch	Rosaceae
49.	Mulberry (Black)	Shehtoot	Morus nigra L.	Moraceae
50.	Mulberry (White)	Shehtoot	Morus alba L.	Moraceae
51.	Olive	Zaitoon	Olea europaea L.	Oleaceae
52.	Papaya	Papita	Carica papaya L.	Caricaceae
53.	Passion fruit	Passion fruit	Passiflora edulis Sims.	Passifloraceae
54.	Phalsa	Phalsa	Grewia asiatica D.C.	Tiliaceae
55.	Pineapple	Arianas	Ananas comosus Merr.	Bromeliaceae
56.	Plum (European)	Alucha	Prunus domestica L.	Rosaceae
57.	Walnut	Akhrot	Juglans regia L.	Juglandaceae
58.	Tamarind	Imli	Tamarindus indica L	Leguminosae
59.	Sapota	Chiku	Achras sapota L.	Sapotaceae
60.	Raspberry	Rasbhari	Rubus idaeus L.	Rosaceae

NOMENCLATURE OF VEGETABLE CROPS

S. No.	English Name	Common Name	Botanical Name	Family
1	Onion	Pyaz	*Allium cepa*	Amaryllidaceae
2.	Garlic	Lahsun	*Alium sativum*	Amaryllidaceae
3.	Leek	Leek	*Alium porum*	Amaryllidaceae
4.	Taro	Arbi	*Colocasia esculenta*	Araceae
5.	Elephant foot yam	Jimikand	*Amorphophyllus sagittifolium*	Araceae
6.	Sweet corn	Sweet corn	*Zea mays*	Graminae
7.	Asparagus	Asparagus	*Asparagus officinalis*	Lillaceae
8.	Amaranthus	Chaulaee	*Amaranthus spp.*	Amaranthaceae
9.	Beet root	Chukander	*Beat vulgaris*	Cheanopodiaceae
10.	Palak	Palak	*Beat vulgaris var bengalensis*	Cheanopodiaceae
11.	Spinach	Vilayati palak	*Spinacea oleracea*	Cheanopodiaceae
12.	Lettuce	Lettuce	*Lactuca sativa*	Compositae
13.	Globe artichoke	Hathiichoke	*Cynara scolymus*	Compositae
14.	Jerusalem artichok	Girasole	*Helianthus tuberosus*	Compositae
15.	Sweet potato	Shakarkand	*Ipomea batatas*	Convolvulaceae
16.	Cabbage	Bandgobi	*Brassica oleracea var. capitata*	Cruciferae
17.	Cauliflower	Fulgobi	*Brassica oleracea var. botrytis*	Cruciferae
18.	Brussel's prout	Mini cabbage	*Brassica oleracea var. gemmifera*	Cruciferae
19	Sprouting Broccoli	Broccoli	*Brassica oleracea var. Italica*	Cruciferae
20.	Kno-khol	Kholrabi	*Brassica oleracea var. gongylodes*	Cruciferae
21.	Kale	Hak sag	*Brassica oleracea var. acephala*	Cruciferae
22.	Chinese abbage		*Brassica chinensis*	Cruciferae
23.	Turnip	Shalgam	*Brassica rapa*	Cruciferae
24.	Radish	Muli	*Raphanus sativus*	Cruciferae
25.	Cucumber	Kheera	*Cucumis sativus*	Cucurbitaceae
26.	Musk Melon	Kharbooj	*Cucumis melo*	Cucurbitaceae
27.	Gherkin	Gherkin	*Cucumis angurai*	Cucurbitaceae
28.	Water melon	Tarbooj	*Citrullus lanatus*	Cucurbitaceae
29.	Round melon	Tinda	*Citrullus lanatus var. fistulosus*	Cucurbitaceae
30.	Pumpkin	Kaddu	*Cucurbita moshchata*	Cucurbitaceae
31.	Summer squash	Chappan kaddu	*Cucurbita pepo*	Cucurbitaceae

S. No.	English Name	Common Name	Botanical Name	Family
32.	Winter squash	Vilayti kaddu	*Cucurbita maxima*	Cucurbitaceae
33.	Bottle gourd	Lauki	*Lagenaria siceraria*	Cucurbitaceae
34.	Bitter gourd	Karela	*Lagenaria siceraria*	Cucurbitaceae
35.	Ridge gourd	Satputia	*Luffa acutangula*	Cucurbitaceae
36.	Sponge gourd	Tauri	*Luffa cylinderca*	Cucurbitaceae
37.	Pointed gourd	parwal	*Trichosanthus anguina*	Cucurbitaceae
38.	Chow-chow	Chayote	*Sechium edule*	Cucurbitaceae
39.	Cassava	Tapioca	*Manihot esculanta*	Euphorbiaceae
40.	Peas	Matar	*Pisum sativum*	Leguminoceae
41.	French bean	Kidney bean	*Phaseolus vugaris*	Leguminoceae
42.	Cluster bean	Guar	*Cyamopsis tetragonolobus*	Leguminoceae
43.	Winged bean	Foxtail bean	*Psophocarpus tetragonolobus*	Leguminoceae
44.	Cow pea	Lobia	*Vigna unguiculata*	Leguminoceae
45.	Fenugreek	Methi	*Trigonella foenu graceum*	Leguminoceae
46.	Okra	Bhindi	*Abelmoschus esculantus*	Malvaceae
47.	Drum stick	Drum stick	*Moringa oleifera*	Moringaceae
48.	Potato	Aalu	*Solanum tuberosum*	Solanaceae
49.	Tomato	Tamatar	*Lycopersicon esculentum*	Solanaceae
50.	Brinjal	Egg plant	*Solanum melongena*	Solanaceae
51.	Sweet pepper	Shimla mirch	*Capsicum annum*	Solanaceae
52.	Chilli	Mirchnu	*Capsicum annum*	Solanaceae
53.	Carrot	Gajar	*Daucus carota*	Umbeliferae
54.	Corinder	Dhaniya	*Coriandrum sativum*	Umbeliferae
55.	Celery	Celery	*Apium graveolens*	Umbeliferae

NOMENCLATURE OF FLOWER CROPS

S. No.	English Name	Common Name	Botanical Name	Family
1	Amaryllis	Belladonna lily	*Amaryllis spp.*	Amarllidaceae
2.	Begonia	Begonia	*Begonia odorata*	Begoniaceae
3.	Canna	Indian Shot	*Canna indica*	Cannaceae
4.	Cyclamen	Florist's cyclamen	*Cyclamen persicum*	Primulaceae
5.	Freesia	Fressia	*Freesia spp.*	Iridacea
6.	Heliconia	Heliconia	*Heliconia spp.*	Helliconiaceae
7.	Iris	Iris	*Iris spp*	Iridaceae
8.	Narcissus	Nargis	*Narcissus tazetta*	Amaryllidaceae
9.	Saffron	Saffron	*Crocus sativus*	Iridaceae
10.	Spider lily	Spider lily	*Hymenocallis littoralis*	Amaryllidaceae

S. No.	English Name	Common Name	Botanical Name	Family
11.	Tulip	Tulip	*Tulipa gesneriana*	Lilaceae
12.	Zantedeschia	Calla lily	*Zentedschia spp.*	Araceae
13.	Zephyranthes	Fairy lily	*Zephyranthes spp.*	Amaryllidaceae
14.	Amaranthus	Love lies bleeding	*Amaranthus caudatus*	Amaranthaceae
15.	Balsam	Gulmehandi	*Impatiens balsamina*	Balsamaceae
16.	Cock's comb	Murgkesh	*Celosia cristata*	Amranthaceae
17.	Gaillardia	Blanket flower	*Gaillardia pulchella*	Compositae
18.	Gomphrena	Globe amaranth	*Gomphrena globosa*	Amaranthaceae
19	Kochia	Bruning brush	*Kochia scoparia*	Chenopodiaceae
20.	Portulaca	Sun plant	*Portulaca grandiflora*	Portilaceae
21.	Sunflower	Surajmukhi	*Helianthus annus*	Compositae
22.	Tithonia	Maxican sunflower	*Tithonia speciosa*	Compositae
23.	Zinnia	Youth and age	*Zinnia elegans*	Compositae
24.	Ageratum	Floss flower	*Ageratum houstonianum*	Compositae
25.	Aster	Ostrich feather	*Callistephus chinensis*	Compositae
26.	Calendula	Pot marigold	*Calendula officinalis*	Compositae
27.	Cineraria	Florist's cineraria	*Senecio cruentus*	Compositae
28.	California poppy	Golden poppy	*Eschscholtizia california*	Papaveraceae
29.	Candy tuft	Hyacinth flower	*Iberis umbellate*	Cruciferaceae
30.	Clarkia	Garland flower	*Clarkia elegans*	Onagraceae
31.	Clianthus	Parrot's bill	*Clianthus dampieri*	Leguminoceae
32.	Coreopsis	Tick seed	*Coreposis tinctoria*	Compositae
33.	Corn flower	Hurt sickle	*Centaurea cyanus*	Compositae
34.	Cosmos	Mexican aster	*Cosmos bipinnatus*	Compositae
35.	Dahlia	Dahlia	*Dahlia variabilis*	Compositae
36.	Daisy	English daisy	*Bellis perennis*	Compositae
37.	Gazania	Treasure flower	*Gazania splendens*	Compositae
38.	Gypsophila	Baby's breath	*Gypsophila elegans*	Caryophyllaceae
39.	Anthurium	Flamingo flower	*Anthurium andreanum*	Araceae
40.	Bird of paradise	Strelitzia	*Strelitzia spp.*	Musaceae
41.	Carnation	Carnation	*Dianthus caryophyllus*	Caryophyllaceae
42	Chrysanthemum	Guldaudi	*Dendranthema grandiflora*	Asteraceae
43.	Crossandra	Firecracker flower	*Crossandra spp.*	Acanthaceae
44.	Dahlia	Dahlia	*Dahlia pinnata*	Asteraceae
45.	Gerbera	Barberton daisy	*Gerbera jamesonii*	Asteraceae
46.	Gladiolus	Sword lily	*Gladiolus grandifloras*	Iridaceae

S. No.	English Name	Common Name	Botanical Name	Family
47.	Jasmine	Chamali	*Jasminum spp.*	Oleaceae
48.	Lily	Lilium	*Lilium spp.*	Liliaceae
49.	Marigold	Gainda	*Tagetes spp.*	Compositae
50.	Orchids	Orchids	*Dendrobium spp.,Vanda spp.*	Orchidaceae
51.	Rose	Rose	*Rosa spp.*	Rosaceae
52.	Tuberose	Rajnigandha	*Polianthes tuberosa L.*	Amaryllidaceae

Exercise No. 3

OBJECTIVE - Preparation of Nursery Bed and Raising Seedlings

NURSERY

Nursery is a place, where seedlings, saplings, trees, shrubs and other plant material are grown and maintained until they are placed in permanent place. Seedlings at early stages deserve special attention which is only possible in nursery. Setting up of a horticultural nursery is a long term venture and requires careful planning and expertise. Establishment, management and marketing are major considerations of nursery.

Establishment of Nursery

The nursery should be established in such an area where cultivation of fruit crops is on sizeable area and there is need for a nursery, having demand for saplings. In such area/region, following considerations need to be observed for selecting an appropriate location.

1. Nursery should be established in the important production areas.
2. Nursery soil should be deep, fertile, well drained and free from pathogens.
3. The site should be well connected by different means of communication and must be easily accessible.
4. The area should be well protected.
5. Soil and micro-climatic conditions should be appropriate.
6. Availability of irrigation and power supply should be ensured.
7. Sufficient labour and skilled person should be available to handle different operation..
8. Availability of progeny/mother stock (bud wood source tree) and root stocks should be ensured.

9. Availability of propagating/growing structure, hardening chamber etc.
10. Availability of working shed, pot and packaging yard and sale unit etc.
11. Availability of office building, staff quarters etc.

Purpose - One of the important operations for horticulture crop is raising of seedlings (fruit trees, vegetables, shrubs and flowering annuals). In many plants, the seeds are first sown in seed beds and after sometimes seedlings are either directly transplanted in the actual field as in case of some vegetables, flowering annuals, shrubs, and trees or may used as root stock as in case of fruit trees like mango, guava, aonla etc. For healthy and good stand of seedlings proper preparation of nursery bed is essential. Since large number of seedlings are raised in small area, due consideration of adequate moisture and nutrients, protection from pest & diseases, strong sunlight and wind is very important. These factors will vary if seedlings are raised in protected structure.

Materials required- Spade, khurpi, rake, rope, pegs, measuring tape, water cane, plastic sheet, coarse sand, seed, manures and fertilizer insecticide, pesticide, fungicide etc.

Procedure- Select the proper area for preparing nursery bed having well drained fertile loamy soil rich in organic matter and sunny situation. Mark out the area required for nursery. Dig the soil about 25-30 cm deep with the help of a spade or kudali. Remove weeds, weed roots and stones etc. Level the land with the help of rake. If the soil is heavy clay or silt then mix sand at the rate of 4-5 kg per sq. metre. After this spread well decomposed FYM (Farm Yard Manure) at the rate of 5-6 kg per sq. metre. Fertilizers like urea, single super phosphate and potassium chloride at the rate of 200-250 g per sq. metre.

Preparation of Nursery bed- Nursery beds of 1- 1.20 m wide and convenient length 5-6 m are considered ideal. In between two parallel beds a space of 50-60 cm should be provide for cultural operations. The height of bed should be 15-20 cm particularly during rainy season and 10-15 cm during winter season from the soil surface.

Type of Nursery Bed

1. **Raised Nursery bed-** During rainy season, it avoids losses through water stagnation by facilitating proper drainage of excess water. The size of nursery bed should be 1.0 m in width and length of nursery bed can be increased upto 5-6 m as per need; Soil is drawn from the sides, so that the level of bed is raised to 15-20 cm from natural level of plot. The soil is worked properlyby raking. All big clods, stones, plant stumps, weed plants etc. are picked up. Well rotten FYM or leaf mould @ 2.5 kg/m^2 mixed. The surface of nursery bed is madesmooth and should be slightly raised in centre with little slope on two sides for draining excess water from middle of the bed. In between two raised beds an alley of about 50-60 cm is left.

2. **Flat Nursery bed-** During spring and summer season seedlings are raised in flat beds. Further, in areas where rainfall is not so heavy and field is well leveled and drained, flat bedis preferred. Inside the bed, surface is leveled and stone, stumps etc. are picked up. Soil is dug at 15-20 cm depth and all clods should be broken with the help of *Kudali* or spade to pulverize the soil.

3. **Sunken Nursery bed-** During summer, seedlings should be raised in sunken type of nursery bed. It protects the plants from hot dry winds. It is made in dry areas. The objective of sunken bed is to avoid flow of water outside the bed. Sunken beds are made by excavating the soil in the bed area. These beds are usually 15 cm deeper than the normal ground level. It is better to connect such sunken beds to a common drainage line so that water does not stand during rains. Major advantage of sunken nursery bed is good water retention for dry conditions and less evaporation.

TYPES OF NURSERY (BASED OF VERSATILITY)

1. **Temporary nursery:** It consists of raised nursery beds. It can be changed from one place to another depending on needs.

2. **Permanent nursery:** Side walls with drainage holes are constructed with concrete to a height of 75 cm. Seeds are sown in soil inside the concrete structure. After removal of each batch of seedlings the soil is enriched with manures.

Precautions

- Prepare the raised beds to avoid damping off. The surface of the beds shouldbe made fine and smooth.
- Provide sufficiently large drainage hole to the seed pans.
- Use porous media to provide good aeration.
- Treat the seeds against seed born diseases before sowing.
- Maintain uniform and proper depth of sowing
- Water the seed beds and seed pans regularly.

Soil treatment/ Sterilization of Nursery beds- Several times, soil contains certain harmful pathogens, which interferes with growth and development of seedlings. These harmful insects can be eliminated by pasteurization, solarization, changing the sight of nursery or by chemical treatment.

Solarization- It is simple and effective method for soil sterilization. For this, soil mulching is done with black polythene of 200-300 gauge during the period of high temperature and solar radiation. The edges of polythene sheet should be pressed in the soil to minimize the air circulation. Soil should be kept moist before mulching to increase the latent heat and thermal sensitivity for resting he soil borne pathogen, harmful insects and weeds which can be reduced to a sustainable level.

CHEMICAL TREATMENT SEED BED TREATMENT

1. Formaldehyde (Formalin) 1% can also be used for soil treatment. 5 litre of this solution is sprayed uniformly on soil surface per sq. meter. After spraying the beds should be covered with black polythene. The edges of the polythene need to be sealed with wet soil to make air tight. Polythene should be removed only after 10-12 days. The seed should be sown only after 6-7 days after removing the polythene.
2. Spray the bavistin 0.1% - 0.3% on the moist bed or Sevin dust 20-25 gm/ sq.m area and leave the bed for 2-3 days.
3. 1% solution of Trichoderma virdi also can be used for seed bed treatment.

Seed treatment- Before sowing the seeds in the sterilized bed, seed should be treated with 0.3% suitable fungicide eg. ceresan, bavistin, thiram etc.

Sowing of seed- In well prepared beds, the seed should be sown in miniature furrows. The depth of furrows will depend upon the seed size. For bolder seeds of furrow should be kept 5-6cm kept and for smaller and fine seed depth should be kept 1-3 cm. The furrows should be covered with a mixture of leaf mould and sand (3:1).

Care and handling of Seedling- The protection should be against strong sun and rain, frost etc. Covering the beds with agro-net/ mosquito net at a height of a meter will protect the seedlings against strong sunshine and wind. Spraying with fungicide and insecticide as and when required proves very useful to the seedling health.

COMPONENTS OF MODERN NURSERY

A number of structures may be necessary for raising a nursery. To establish the nursery, the following structures need to be constructed:

1. **Fencing:** It is required to protect a nursery particularly from stray animals.
2. **Progeny block (bud wood source tree) :** In progeny block, true to type mother plant are maintained in the nursery. Suitable plant types with existing superior cultivars/varieties should be collected and maintained in the progeny block. The mother plants should be true to type, healthy, heavy bearer with standard quality fruit having tolerance to biotic and abiotic stresses. Care should be taken to label the plants properly.
3. **Rootstocks and seed gardens-** After establishment of scion bank, next priority should be given for the establishment of the rootstocks or seed garden. Seed propagation is the most usual way for mass production of rootstocks.
4. Growing structures: There should be provision of modern propagation structure like greenhouse/polyhouse, mist chamber etc. these structure provide optimum growing conditions for seed germination, rooting of cutting, hardening of seedlings. Now a day's green house has become a prerequisite of Hi-Tech nursery. Shade nets are useful not only for reducing heat injury to young plants, but their use also reduces transpiration. Shade nets are available in different colours and densities.

5. High Humidity Chamber: This technique resolves the common problem of grafts or cuttings dying due to desiccation (drying up) when planted in the soil for rooting, by ensuring a humid atmosphere around the cuttings, thus preventing excessive evaporation.
6. Use of Supplementary Light: Several plants go into winter dormancy when the day length gets short. Additional light from tube lights, given after sunset, creates long-day condition that prevents the plants from going into winter dormancy.
7. Drought Hardening: Plants that are raised under high atmospheric humidity and shade often die due to transplanting shock when shifted to the fields. To prevent this, the plants are hardened by allowing external dry air to enter the chamber gradually.

After having established the above infrastructure, the nursery establishment and planning involve division of the nursery into different units, viz., propagation unit, production unit, packaging unit and sale unit.

1. Propagation Unit

Propagation unit is the major unit of nursery work and includes:

i. Actual propagation structures such as green house, hot beds, cold frames and mist house.
ii. Service structures such as head house.
iii. The alley house connecting to the hot beds and cold frames with head house. It provide a passage for the transport of plants, propagation media, soil and fertilizers from head house to propagation structures, and must be sufficient wide to permit easy and quick movement.

i. Primary Nursery (Seedbeds)

Seed beds near to water source and to office so that they can be kept under vigilant control. The raised seedbeds of 6-8 cm height, 1 meter width and of convenient length, free from stones should be prepared with upper 2.5-5cm of the bed filled with sand. Soil can be prepared to fine tilth, add sufficient quantity of rotted FYM, vermin-manure or pig manure at least 10-15 days earlier of seed sowing. The bed may be treated with 1% Bordeaux or 0.1% Bavistin before sowing of seeds.

ii. Nursery Beds

Seedlings from seed beds are removed and transplanted in the nursery beds. Nursery beds should be located in an open area near to water source. Nursery beds should be prepared by adding sufficient organic manures and fertilizers. Nursery beds should be divided into section as per crop and varieties. The nursery beds should be laid out in such a way that there is an access to all the beds through roads or paths.

iii. Pot Yard

The pot yard should be in shade because the tender plants require shade as compared to hardy plants. This section should be near to water source. Trenches can be provided for keeping potted plants closely packed together.

1. **Production Unit :** The object of this unit is to rear the new plants from seedling to marketable stage. This unit is divided into different blocks, each block being meant for only one kind of plant or species. This helps on sorting of plants, easy record keeping and doing the operation as per the need of a plant species.

2. **Packaging Unit:**The packing yard is used for packing the plants before sale or dispatch to out stations. The yard can be combined with working shed. It is near to sale counter.

3. **Sale Unit:** The objective of the sale unit is to market the nursery plants effectively. The design and layout of this unit should be attractive to the customers. This should be usually located on a well travelled way and may be by the side of production unit. The sale unit is usually divided into three different parts such as display unit, sale and packing, and parking unit.

Raised Bed

Flat Bed

Sunken bed

Exercise No. 4

OBJECTIVE – Preparation of Seedlings in Protrays

Material required: Cocopeat, perlite, vermiculite, NPK grade water soluble fertilizers,rose can, pro-trays, iron stand, mulching material, seed etc.

Principle: Seedlings raised in portrays can be shifted to field along with intact roots and no any exposure to roots to environment hence seedlings of all the vegetables can be raised and transplanted, even for the crops which are not able to resist transplantingshock i.e. cucurbits, cole crops, tomato, brinjal etc. The growing media used for filling portrays contains vermiculite, cocopeat and perlite in certain proportion. Here cocopeat act as binding media agent and source of organic matter, vermiculite increaseswater holding capacity of growing media as it is expanding type of clay material, where as perlite improves aeration and maintain temperature of growing media. Thereare minimum chances of soil borne diseases and weed infestation as entire growing media and protays are sterilized and free from seed of weed.

ADVANTAGE

a. Seedlings can be raised for those crops also which cannot be raised traditionally including cucurbits, cole crops and beans.

b. Uniform germination and fast growth of young plants due to controlled condition of growing media and microclimate.

c. Distinct transportation becomes possible.

d. Vigorous root growth with more number of roots.

e. Best way for taking early crops of cucurbits because in month of March-April fields are occupied with Rabi season crops and after harvesting of these cropsif sowing will be done then it will take at least 50 days to start harvesting.Here if sowing of cucurbits have been done in pro trays in month of February- March, then one month old seedlings can be transplanted in the field even on the same day of harvesting and produce can be send to the market at least one month earlier.

f. At the transplanting, fast crop establishment without mortality because no any transplanting shock will be there due to intact roots.
g. Seedlings can be raised round the year as per need due to less dependency onthe environment.
h. No any chances of soil borne fungal diseases and other threat due to contaminated soil as soil is not involved in this technique.
i. No any weed will be there for competing with seedlings.

Disadvantage

i. Initial cost may be little bit higher.
ii. More attention and care is required as compare to traditional method.
iii. Trained personal or skilled labour will be required.

Procedure

i. Well cleaned and sterilized pro-trays should be taken for nursery raising, trays can be sterilized in sunlight for 2-3 hours or can be treated 2% formaldehyde.
ii. Take cocopeat, perlite and vermiculite in the ratio of 3:1:1, respectively, on volume basis and mix it well.
iii. Add water in this mixture and thoroughly mix it. Quantity of water must be adequate, means if you press the growing media in palm, it must be bound. Media will not be bound in shape if more or less water is there.
iv. Fill the portrays properly and prepare a minute pit in center of cavity with the help of finger. Depth of pit should not more than 1 cm.
v. Put single seed in the pit and apply growing media to fill trays again, followed by light irrigation with the help of rose can.
vi. After a light irrigation a polythene sheet or thermocol sheet or paper shouldbe place on the pro-trays for better and uniform germination.
vii. Arrange these trays on iron or wooden stand having a height of at least 2-3 feet above the ground so that pro trays will be at working height and roots, cannot be enter in to the soil.
viii. When germination takes place or seedling like irrigation, fertigation should be followed till seedling become ready for transplanting.
ix. After care and management of seedling like irrigation, fertigation should befollowed till seeding become ready for transplanting.
x. Seedlings became ready in 20-40 days according to species and climatic condition.

Precautions

i. All the inputs must be clean and sterilized.

ii. Irrigate seedlings at least once in a day.

iii. Hose pipe with pressure or jet should not used for irrigation.

iv. Entire set up of nursery must be located under controlled condition oftemperature and humidity.

Exercise No. 5

OBJECTIVE – To Study About Growing Media

Definition: Growing medium often referred to as "substrate" or potting soil" is a material, otherthan soil on the spot, in which plants are grown.

FUNCTION OF GROWING MEDIA

In a plant nursery, a growing medium serves four functions:

1. It physically supports the plant.
2. Large pores promote oxygen exchange for root respiration.
3. Small pores hold water
4. Mineral nutrients are carried in water to plant roots

CHARACTERISTICS OF AN IDEAL GROWING MEDIA

A. Physical properties: water holding capacity, aeration, porosity, shrinking and swelling, bulk density.

B. Chemical properties: Fertility, 5.5 and 6.5 pH, Cation exchange capacity.

In polybags for plants/ seeding: Sand25%: Soil 50%: Compost 25%

For Cutting: Sand 50%, soil 25% and compost 25%

For indoor potting: gravel at bottom then sand and above with Cocopeat

For vegetable seedling: Cocopeat: vermiculite : perlite (3:1:1)

Some common growing media for raising the plants are:

a. Soil Mixture

This is the most commonly employed medium for pot plants. It usually consists of red earth, well decomposed cattle manure, leaf mold, river sand and also charcoal in some cases. Soil mixture commonly used for propagation is:

Red earth	-	2 part
FYM	-	1 part
Sand	-	1 part

b. Sand

It is the most satisfactory medium for rooting of cuttings.

c. Compost

It is decomposed and rotten material of farm waste. For the growing seeds and cutting, it is very common and useful material. Compost is rich in organic matter and adds to the growth and vigour of the plants. Mostly compost is prepared using dung which contains faecal passed out, living seed of weeds and may be infested with soil dwelling pests like beetles, grubs and root rotting pathogens. It should preferably be used after sterilization.

d. Peat

It consists of the remains of aquatic marsh, bog or swamp vegetation whichhas been preserved under water in a partially decomposed state. When such peat is derived from sphagnum or other mosses, it is known as peat moss. It is used inmixture after breaking them and moistened.

e. Sphagnum Moss

Commercial sphagnum moss is the dehydrated young residue or living portion of acid-bog plants in the genus Sphagnum such as *S. papilliosum*, *S. capillacem* and *S. palustre*. It is generally collected from the tree trunks of the forest species in south Indian hills above 1500mts. above M.S.L. during rainy period. It is relatively sterile, light in weight and has a very high water-holding capacity. It is the commonly used medium in air layering.

f. Vermiculite

It is very light in weight and able to absorb large quantities of water. This can be used as a rooting medium for air layering and also in pots for raising certain plants.

g. Saw Dust

It is obtained as a waste material from timber industry. It is fine in appearance, holds sufficient moisture and retains moisture for sufficient period. Saw dust takes long time to decompose and convert into organic compound. Therefore, aeration is hampered with the use of saw-dust. It has no nutrient content and hence nutrient should be supplied externally to support the growth of the plants. Being organic in nature, saw dust is ideal for the growth of fungus and hence its use is limited in propagation.

h. Formaldehyde

Formaldehyde is a good fungicide used for fumigating the soil. It is effective in killing micro-organisms and weeds seeds. A mixture of one-gallon (3.8 litre) 3f the commercial formalin (37 per cent strength) with 50 gallons (190 litre) of water.

Applied at the rate of 21 to 42 litre per square metre. The treated soil should be covered immediately with polythene sheet or any other similar material for 24 hours. Covering prevents emancipation of fumes and it penetrates in soil fully. After 24 hours, the covering material is removed and the soil is left exposed for about two weeks. The soil is well dried and aerated to have a complete freeness of odour of formalin. After complete disappearance of fumes, the soil is used for seed sowing or planting.

i. Vapam

Vapam (Sodium N-methyl dithiocarbonate dihydrate) is a water-soluble fumigant. It kills weeds, germinating weed seeds, most soil fungi and nematodes. It is applied by sprinkling on the soil surface, through irrigation system. For seed bed fumigation, 0.95 litre of the liquid formulation of vapam is used with 7.6 to 11.4 litres of water. After application, the vapam is sealed with additional water. After three weeks, the soil can be used for seed sowing.

j. Soil Drenching

Wetting of soil particles till full saturation of soil constitutes soil drenching. It is done to inhibit the growth of soil borne fungi. Pythium, Phytophthora, Fusarium and Rhizoctonia species of fungus cause damping off disease in nursery. Drenching using Thiram, Captain, Ceresan etc. is very effective against damping off disease.

SOME COMMON GROWING MEDIA

Soil

Sphagnum

Sand

Vermiculite

Perlite

FYM

Cocopeat	**Compost**	**Sawdust**

Exercise No. 6

OBJECTIVE - Study of Pots, Potting, Depotting and Repotting

Pots are containers in which seeds are sown, seedlings are raised or plants are maintained. Pot culture is the growing of plants in pots.

Purpose - Potting is one of the important horticultural operations for raising of seed propagated plants eg. Papaya, Acid lime etc. in polythene bag for direct sale and raising of seedling for rootstock purpose. One can also enhance the beauty of interiors by artistic arrangement of potted plants, especially the blooming seasonal, perennials and decorative foliage plants. A terrace or roof garden can be developed by growing fruit trees, shrubs, creepers, cacti, succulents and even vegetables in posts as well as large size tubs. The major limitation is that the space and the quantity of soil are limited in pots; as such growth of plants is restricted.

1. **Types of Pot-**Pots may be classified into following groups -
2. **On the basis of material used**- Earthen (clay) pots, Metallic pots, Plastic pots, Cemented Pots, Fibre pots, Ceramic Pots, Glass pots and Polythene bag, Portrays etc. Among them polythene bag and earthen pots are more common. Flower pot is the relative term used for all the containers.
3. **On the basis of Shape**- Conical, Square, Rectangular, Circular, Cylindrical, Bowl shape etc.
4. **On the basis of size**- Large, medium large, medium, medium small and small size.
5. On the basis of Colour- Green, Yellow, White, Red and multiple colour with different design.

Qualities of an ideal pot: It must have sufficient space along with holes for drainage and fulfill the purpose with desired shape and colour. Potted plants can be easily handled and shifted conveniently to any place for decoration purposes and as per requirements.

Potting mixture – Soil : Sand : FYM/ Vermicompost (1:1:1) enriched with or without Bio fertilizer and plant growth promoting substances

Potting: Generally, potting refers to first planting of seedling or a cutting in a container. It is a process of planting a new plant in pot with a suitable pot mixture for establishment. Although it is a simple operation, it requires certain degree of skill and practice.

PROCEDURE FOR POTTING OF EARTHEN POT

1. Select a good quality earthen pot and immerse in water for about one hour. If old pots are used, thorough cleaning is necessary.
2. Place a good crock on the drainage hole with its concave side facing the hole. Over this a large number of pot pieces are put (4-5 cm thick). On these crocks a layer of coarse sand or gravel or coconut fibre or sphagnum moss is spread to ensure adequate drainage and prevent clogging of drainage hole.
3. Remaining space in the pot is filled with suitable pot mixture leaving a head space of 2.5-5.0 cm.
4. The pot mixture should be sufficiently moist at the time of planting.
5. The plant is placed at the centre by scooping out the required amount of pot mixture, so that it accommodate the root system and held in position by packing the soil mixture with hand gently.
6. Water the pots immediately after planting.

Depotting

The process of separation of plants and soil mixture from the potis called depotting. The objective of depotting is to grow plant in field, to re- fresh the soil mixture and shifting of plants in new pots etc.

Steps during depotting:

- Lightly irrigate the pot so earthen ball bind properly with the roots.
- Left hand's fingers keep over the soil near the base of the plant, turning thepot upside down and tapping its rim.
- All mass of soil with single lump remove from pot and shifted that plant atdesired site with least disturbance of soil ball.
- Depotting plants should be placed under shade and watering should begiven frequently (morning and evening) to avoid wilting.

Repotting

It is generally referred to the transfer of a plant from one pot to another and replacing the soil mixture with the fresh one. The first step in repotting is the depotting. After depotting the plant with compact roots, with mother soil removed from it is placed in the centre of the new pot and then the sides are packed with the new garden mixture.

Repotting is necessary when, nutrients are exhausted, pots are broken due to wind or mechanical damage, the soil turns sourest due to continuous watering, under pot bound conditions, insect and disease infestation and for exhibition purpose.

Potting

Repotting

Exercise No. 7

OBJECTIVE - Study of Orchard Layout

Layout is done to locate the actual position of the trees, roads & sub-road, irrigation & drainage channels, store and buildings etc in the orchard. The mistakes committed in the initial stage of orchard establishment can cause loss throughout the life of an orchard, and it is very difficult or even impossible to correct them later. It is, therefore, essential that the layout of an orchard should be carefully planned and executed to facilitate proper care of the orchard.

Principles- A well considered layout plan should be followed for planting an orchard. The plan should provide optimum number of trees per unit area with sufficient space for the proper development of each tree and convenience in various orchard operations such as inter-culture, spray and harvesting etc.

Materials required- Rope, poles, pegs, Carpenters triangle or cross staff, measuring tape and planting board are required for laying out an orchard.

PROCEDURE FOR LAYOUT STEPS

1. First and foremost step in orchard layout is to draw the base line parallel to the road or fence or the boundary of the orchard. This should be drawn at a distance of half the spacing that is to be followed, for example, if the spacing is 10 metre the base line should be drawn at a distance of 5 metre from the periphery of the plot.
2. Towards the end of the base line, leave again a space equal to half the spacing from the boundary or road or fence etc. and put the peg on one end of the base line. From this peg, measure the planting distance and put the second peg on the base line. Thus, continue placing pegs at each of the planting distances till the total length of the base line is covered. The distance from the last peg to the boundary should also be half of the spacing given.
3. From the first peg and the last peg on the base line, draw perpendicular lines to the base line. The perpendicular lines may be drawn by adopting any of the following methods.

A. Pythagoras theorem or Carpenters triangle: In this case, by adopting a right angled triangle with the sides and hypotenuse in the proportions of 3:4:5, a perpendicular line can be drawn.

Pythagoras principle-In a right angle triangle

(Length of Hypotenuse)^2 = (Length of base)^2 + (Length of Perpendicular)^2
$(5)^2 = (3)^2 + (4)^2$

$25 = 9+16$

Steps:

Thus a triangle with 3 meter base, 4 meter perpendicular and 5 meter hypotenuse will be a right angle.

1. On the base line, mark the point A by a peg and place the point zero of measuring tape at point A. Measure and mark point B using a peg at 3 meter from the point A, with the mark of 3 meter of measuring tape on point B (AB=3m).
2. Keep the mark zero of tape at point A and 3 m at point B, mark point C at 5 m from point B (BC=5m). At present the mark zero of tape is on point A, mark 3m at point B (AB=3m) and mark 4m on point C (CA=4m).
3. It is possible that CA of 4m length is not coming at point A; adjust it by moving point C or point B. Thus, it becomes right angle triangle.
4. Now extend a straight line from the position of the first peg by increasing the length of AB and AC. This gives a perpendicular line to the base line from the position of the first peg.

B. Bilateral or Isosceles triangle: In a bilateral triangle, the line bisecting the base will be perpendicular to the base of the triangle. The principle of Pythagoras theorem requires a scale to measure the distance in the proportion of 3:4:5. But by the bilateral triangle principle perpendicular lines can be drawn very easily even in the absence of any scale to measure the distances.

C. Cross staff: Cross staff comprises of a wooden block with two perpendicular slits made on its surface and fixed on an iron rod.

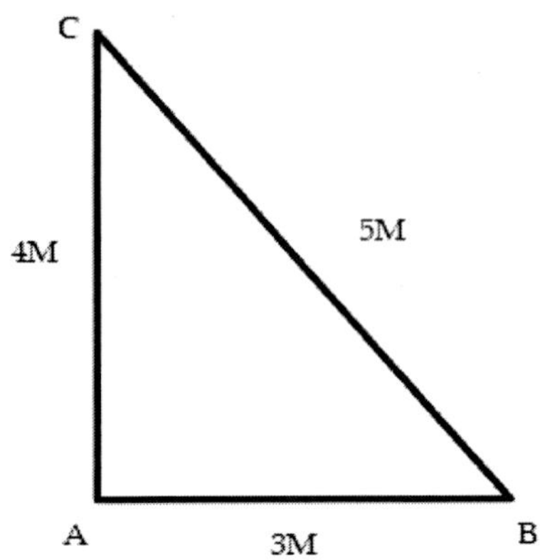

Exercise No. 8

OBJECTIVE – Study about Orchard Planting Systems

Orchard is a piece of land cultivated with fruit crops and related horticultural crops. Establishment of an orchard is a long-term investment and hence needs through planning on selection of site & location, nature of soil & sub-soil, protection of area, layout and procurement of genuine saplings.

There are certain factors which decide the planting distance or spacing:

- **Kind of fruit tree:** Mango (10 x 10 m), guava (5m x 5 m) whereas papaya (2m x 2m) spacing.
- **Rainfall:** In low rainfall areas wider spacing should be provided than high rainfall area.
- **Soil type and soil fertility:** In heavy soil less spacing should be given becausethe top and root growth are limited.
- **Root stocks:** Tree of some variety grafted on different rootstocks will grow to different size and such trees require different planting distance example Apple
- **Pruning and Training:** Trees trained on head system closer spacing than the other type of training.
- **Irrigation system:** If the spacing between the trees is too wide, the yield per unit area would be greatly reduced and vice versa close planting results in a greater yield per unit area in the early life but less in the later years.

Before a grower selects a site for establishing a new orchard, he must have assessed thefollowing factors:

1. Suitability of soil, its fertility, the nature of subsoil and soil depth.
2. During the rainy season, the site must have adequate drainage and no water stagnation.
3. Water for irrigation must be of good quality.

4. There must be adequate transportation options, whether by road or rail, within easy reach.
5. There should be certain consumer demand for the fruits that will be grown
6. Workforce availability

PLANNING OF AN ORCHARD

For the most productive and cost-effective management, the orchard must be meticulously planned. When putting together the strategy, keep the following points in mind.

7. Maximum number of trees per unit area with optimum spacing.
8. In order to ensure adequate supervision, stores and office buildings should be built in the orchard's middle.
9. At a rate of one well per 2 to 4 hectares, wells should be positioned in convenient locationsthroughout the region.
10. Each fruit type should be allocated to its own block.
11. For the sake of transportation economy, roads can take up the least amount of space possible.
12. The space between the windbreak and the first row of trees is beneficial to the lane.
13. Short growing trees should be planted in the front, and tall trees should be planted in the back to allow for easier visibility and to enhance the appearance.
14. Evergreen trees should be in the foreground, and deciduous trees should be in the background.
15. Fruits that attract birds and animals should be planted near the watchman's hut.
16. A good fence is necessary. Compared to other types of fences, live fencing is cost-effective.
17. Drought-resistant, easy-to-produce from seed, fast-growing plants with thick foliage that can withstand extreme pruning and thorny plants are perfect for live fencing. Agave, *Prosopis juliflora, Pithecolobium dulce,* and Thevetia would make a good live fence if planted in three rows.
18. Wind breaks, rows of tall trees planted close together around the orchard, are essential to resist velocity of wind which causes severe ill-effects particularly moisture evaporation from the soil. Since the wind breaks are very effective in reducing the wind velocity and minimizing the damage to the fruit trees and to other crops, their presence in regions where strong winds prevail is of paramount importance. A wind break ordinarily has its maximum effectiveness for a distance about four times as great as its height but has some effect over twice about that distance. Trees suitable for windbreak should be erect, tall and quick growing, hardy and drought

resistant and mechanically strong and dense to offer maximum resistance to wind. The trees which are suitable for growing as wind breaks are Casuarina equisetifolia, Pterospermum acerifolium, Polyalthia longifolia, Eucalyptus globulus, Grevillea robusta, Azadirachta indica.

COMMONLY FOLLOWED SPACING OF FRUIT CROPS

Crop	Spacing (Row x Plant)
Pineapple	30 cm x 60 cm x 90 cm
Banana, papaya and grape	1.8-2 m x 3 m
Passion fruit, phalsa	2 m x 3 m, 3 m x 3m
Custard apple and pomegranate	4-5 m x 4-5 m
Peach, fig, pomegranate, guava	5 m x 5 m
Mandarin, lime, lemon, sweet orange, almond, mulberry, guava	6 m x 6 m
Pumelo and grapefruit	6-7 m x 6-7 m
Date palm, bael, litchi, ber	8 m x 8 m
Sapota, loquat, avocado	8-9 m x 8-9 m
Mango, walnut, cherry	10 m x 10 m
Anola, walnut, cherry	9-11 m x 9-11 m
Jamun, tamarind, pecannut, jackfruit, bread fruit	12 m x 12 m
Chestnut	15 m x 15 m

SYSTEM OF LAYOUT

System or methods of plant layout- The plan showing the arrangement of plant in an orchard is known as plant layout. Although several systems of planting are followed, but selection of a suitable system, depending on soil, climate, plant type, system of training and pruning is very important. Adoption of improper system results in over lapping of plant parts and competition for water, light, nutrient and unequal distribution of water etc

There are several planting plans or systems which can be adopted for planting an orchard. The different system of planting is as follows:

1. Square System
2. Rectangular System
3. Triangular System
4. Quincunx System
5. Hexagonal System
6. Contour/ Terrace System
7. Hedge System

I. Square System

This is the simplest of all systems. In this system of planting the plants are planted in straight rows running at right angle. The distance between plants and between rows remains same. Plants are planted at the corners of a square. Better watching and the possibility of cultural operations in two directions is the greatest advantage of this system. The major disadvantage of this system is that a lot of space is wasted in between the squares.

II. Rectangular System

This system is similar to that of the square system in its layout except for the difference that the spacing between the rows and between the plants in a row is not equal. In this system, trees are planted on each corner of a rectangle. The wider alley spaces available between rows trees permit easy intercultural operations and even the use of mechanical operations. The major disadvantage in this system is that two way inter cultivation is not possible.

III. Triangular System

This system is similar to the square system of planting except that in every alternate row the plants are planted in midway of two plants of the previous row. Thus, tree plants make a triangle where only two arms are of equal length.

IV. Quincunx System

This is also known as filler or diagonal system. This system is essentially the square system except for an additional tree in the centre of each square. Thus the number of trees are nearly double than the square system, but does not provide equal spacing. Center (filler) trees may be short lived. This is difficult layout on ground and can be adopted when spacing for permanent tree is more than 10m. This is not satisfactory as a permanent plant but is satisfactory for putting temporary trees in the centre of squares. Filler should be removed after a few years when main trees come to bearing.

V. Hexagonal System

This system is also known as equilateral triangle system of planting. This system is also called septuplet because seventh tree is put in the centre of the hexagon. The plant in this system is planted at the corners of the equilateral triangle with one tree in the centre. Thus, six trees make a hexagon with an additional tree in the centre of the hexagon. The perpendicular distance between any two adjacent rows is equal to the product of 0.866 x the distance between any two trees. As the perpendicular distance between any two rows is less than unity this system allows 15% more plants than the square system. The limitation of this system is that it is difficult to lay out and the inter cultivation is not so easily done as in the square system. Besides, the watch and ward also becomes difficult as one cannot see in all the directions from a point.

VI. Contour System

This system of planting is usually followed on hills with high slopes. The layout is started from the lowest level and the tree rows are planted along a uniform slope, at right angles to the slope, with a view to reduce loss of top soil due to erosion. This is necessary for rolling topography. Trees can be planted on terraces or along contours. Terraced fields rise in steps one above the other and help to bring more area into productive use and also to prevent soil erosion. In South India, tea is planted in contours either in single hedge system or in double hedge system. Double hedge contour planting system accommodates nearly 22% higher population than single hedge system.

VII. Hedge System

The layout is exactly same as rectangular system except that very wider spacing is maintained between rows and a very narrow spacing is followed between plants. This system permits easy movement of men, material and machinery and also effective cultural operations due to wider spacing. Therefore, this system is especially suitable where machines are employed for various farm operations.

High density planting: An approach where accommodate more numberof plants over conventional planting density for higher productivity and utilization of resources in better wary. In India, the usefulness/vitality of this technology has been proved in an array of fruit crops e.g., pineapple, banana, papaya, guava, mango, apple and citrus.

Comparison of productivity under HDP system with average nationalproductivity of some Indian fruits

Crop	Cultivar	Spacing (m)	Pant Population (ha-1)	Yield under HDP (MT ha-1)	Av. National Yield (MT ha-1)
Banana	Dwarf Cavendish	1.5 x 1.5	4444	120	34.4
Papaya	Coorg Honey Dew	1.2 x 1.8	4262	146	40.9
	Pusa Nanha	1.25 x 1.25	6000	155	46.4
Mango	Amarpali	2.5 x 2.5	1600	22 (10 years old)	6.5
	Dashehari	3 x 2.5	1333	34.74 (24 year old)	8.4
Guava	Allahabad Safeda (Root Stock, Aneupliod-82, Pusa Srijan)	3 x 2	1111	20.4	11.7
Pineapple	Kew & Queen	0.25-0.30 (Plant) 0.45-0.60 (row) 0.90-1.05 (bed)	43000-63750	60-70	15.1

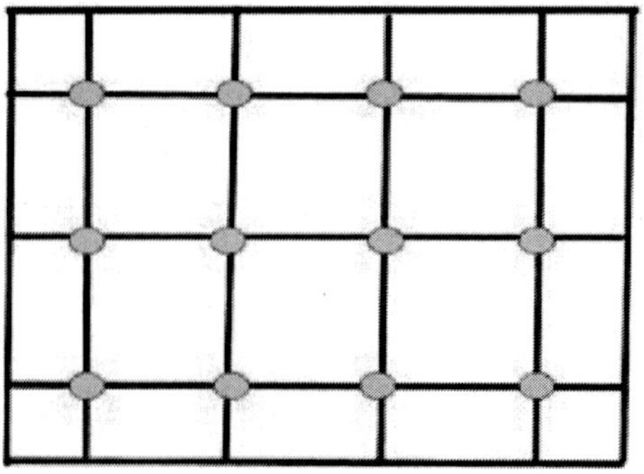

Square

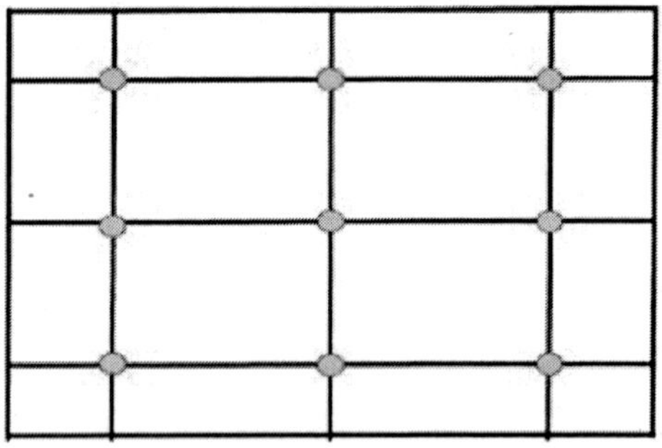

Rectangular

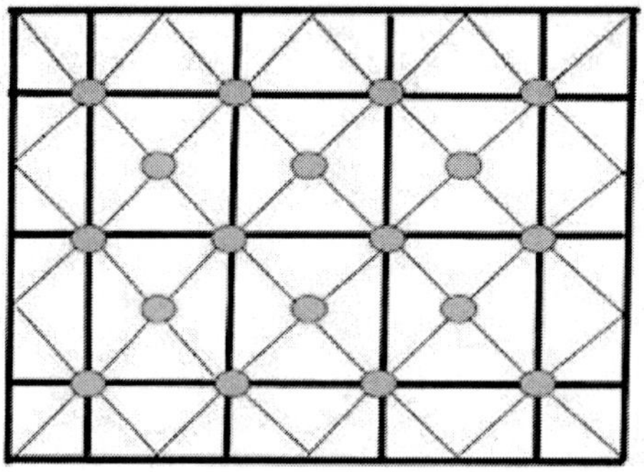

Diagonal/Quincunx

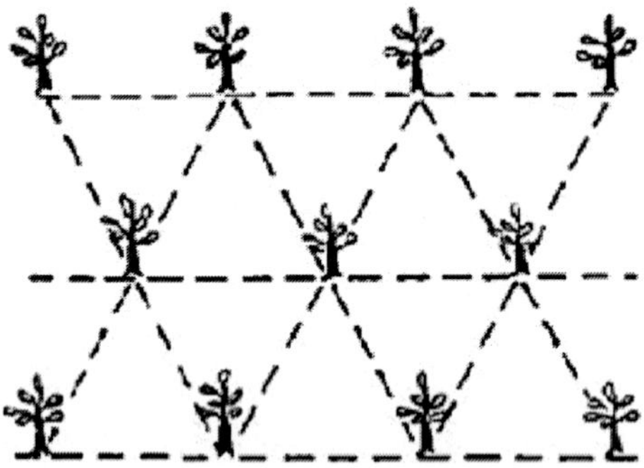

Triangular System

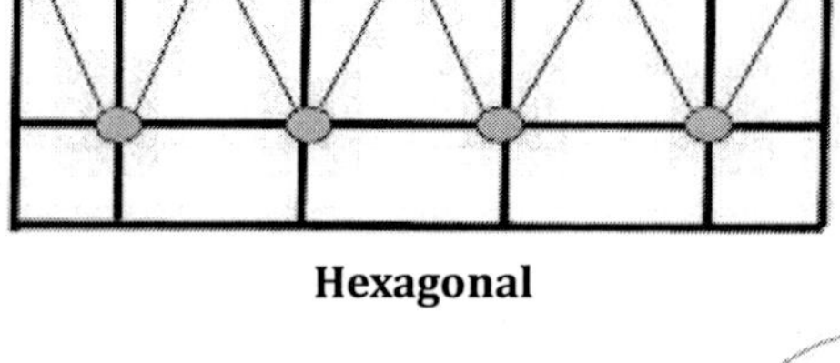

Hexagonal

Terrace System

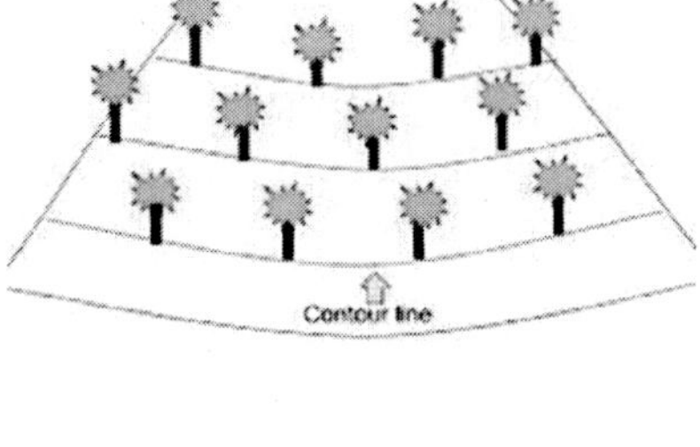

CALCULATION OF NUMBER OF PLANTS IN DIFFERENT SYSTEMS OF PLANTING

The number of plants that can be accommodated by each of the systems in a unit area should be calculated by the formula shown against each system as under:

a. Square System $= \frac{A}{LxP}$

A= Field Area

L= Row to Row spacing

P= Plant to Plant spacing

Example- Area is 10000 sq. metre and planting distance is 10x10 (m) then

Number of Plants $\frac{10,000}{10 \times 10} = 100$ Plant

b. Rectangular System $= \frac{A}{LxP}$

A=Field Area

L= Row to Row spacing

P= Plant to Plant spacing

Example- Area is 10000 sq. metre and planting distance is 10x 8 (m) then- Number of plants=10,000/10 8 = 125 Plants

c. Quincunx System

As the plants are planted additionally in the centre of the square, hence first the number of plants is calculated for square system of planting which is-

No of Plants= Area in square metre/Planting distance in metre square=10,000/10 X 10 = 100 Plants.

Additional plants = (No. of rows length wise - 1) (No. of rows width wise -1)

In 100 100 sq. metre field if planting distance is10 10 m. then number of rows length wise and width wise will be 10

Hence, No of additional plants (10-1) (10-1) = 9 9 = 81

Total number of plants = Pants planted in Square system of planting + additionally planted plants in the centre of square ie. 100 + 81 = 181

d. Hexagonal system $= \frac{\text{Area} \quad \times 115}{\text{Spacing } 100}$

e.Triangular system $= \frac{S}{D^2 \times 0.8666}$

S= unit surface

D = Length of the triangle side

Exercise No. 9

OBJECTIVE - Propagation Through Specialized Vegetative Structures

Propagation-The multiplication of plant through seed or vegetative means is known as propagation. There are two methods of producing new plants-

1. **Sexual method-** The multiplication of plant through seed is known as sexual method of propagation. Fertilization is a sexual reproduction process in plants that occurs after pollination and germination. The fusion of male gametes (pollen) and female gametes (ovum) to form a diploid zygote is known as fertilization. It is a physicochemical reaction that happens after the carpel has been pollinated. The entire sequence of this process occurs in the zygote as it develops into a seed. Flowers, as the reproductive structures of angiosperms, play an important role in the fertilization process (flowering plants). When gametes in haploid conditions fuse to create a diploid zygote, this is the method of fertilization in plants. Male gametes are transferred into female reproductive organs by pollinators (honey bees, birds, bats, butterflies, flower beetles) during fertilization, and the end result is the development of an embryo in a seed.

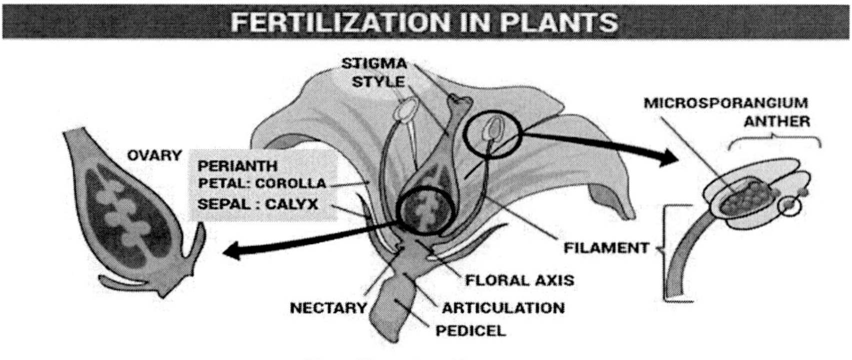

Fertilization Process

2. **Asexual or vegetative method-** The multiplication of plant through vegetative means is known as asexual method of propagation. It may be through division or separation/division, cutting, layering, budding and grafting.

There are certain plant modifications which are used for vegetative propagation of plants. These modified plant parts may be stem, root, or leaves and are usually specialized for food storage. Two principal methods are used for propagation of plants by using these modifications.

A- Separation: naturally detachable structures, such as bulbs or corms are separated and planted individually and

B- Division: The plants modification such as rhizomes, tubers etc., are cut into sections to obtain new plants from each section.

1. **Bulbs:** Bulbs are produced by monocotyledonous plants in which the stem is modified for storage and reproduction. Bulb is a specialized underground organ consisting of a short freshly, usually vertical stem axis bearing at tip apex or growing points and enclosed by thick freshly scales. Bulb scales morphologically are the continuous sheathing leaf base. Growing points develop in the axils of these scales to produce miniature bulbs known as bulbets/ daughter bulbs. These daughter bulbs cane separated from the mother plant at the end of growing season and used as propagating material.

   Ex: Tulip, Daffodils, lilum, Tuberose, Onion, Garlic (cloves)

2. **Tubers:** A tuber is the short terminal portion of an underground stem which has become thickened because of accumulation of preserved food material eg: Potato, Jerusalem artichoke, yam. Propagation by tuber can be carried out either by planting the whole tuber or by cutting into sections each containing bud or eyes.

3. **Tuberous roots:** Certain herbaceous perennials produce thickened roots which contain large amount of stored food. The tuberous roots differ from the tubers in that they lack nodes and internodes. Adventitious buds are present only at stem end or proximal end; fibrous roots are produced towards the distal end. These fleshy roots are separated and used for propagation. For example- Sweet potato, Dhalia, Tapioca (Cassava).

4. **Rhizomes:** The horizontal, thick and fleshy or slender and elongated stem growing underground are known as rhizomes. Rhizomes have nodes and internodes and readily produce adventitious roots. The rhizomes are cut into pieces, each containing vegetative bud and transplanted. Eg: Banana, Ginger, Ferns, Turmeric, and Cardamom.

5. **Corms:** A corm is solid underground base of a stem having nodes and internodes and is enclosed by a dry scale like leaves. After flowering one or more corms may develop just above the old one, which disintegrates. In addition several new corms called caramels develop below each new corm. These may be separated and grown for 1-2 years to reach flowering stage. Eg: Gladiolus, Amor phophallus, saffron.

6. **Runners**: Runners are specialized arial stems (stolons) arising in the leaf axils of plant having rosette crowns. New plants arise from nodes at interval along these runners. From these runners more new runners may arise thus developing natural clonal multiplication methods. The typical runner producing plant is straw berry which is photo sensitive with regard to its runner production. Long days favour runner production where as short days prevent runner formation. Eg: Strawberry, mint.

7. **Suckers**: Adventitious shoot from the underground portion of the stem or from their horizontal root systems are known as suckers and when these strike roots, they may be utilized as propagation materials. Well developed suckers are dugout and separated from the mother plant and planted in the nursery for further growth. Suckers are usually treated like rooted layers. Eg: Pineapple, Chrysanthemum, Curry leaf, Banana.

8. **Offsets/ offshoots:** An offset is a shoot or thick stem of rosette like appearance arising from the base of the main stem of certain plant such as date palm, pineapple etc., Date palm cultivars are propagated vegetatively by separating away the offshoots and replanting them. However these are girdled and layered for about a year prior to separation, because offshoots do not root easily when directly separated from the mother plant and planted in the field.

9. **Stolon:** A stolen is an aerial shot that hits roots after colliding with the earth. It may be a prostate or a horizontally rising stem from the crown. *Cyanodon dactylon,* for example. A shoot that has been rooted this way is simply cut from the parent plants and transplanted or potted.

10. **Bulbils:** Some lily varieties, such as Lilium bulbiferum and Lilium tiginum, produce aerial stem bulblets, also known as bulbils, in the axils of their leaves. Bulbils bloom early in the season and fall to the ground several weeks after the plant has finished flowering. They are harvested and planted shortly before they fall naturally. When flower buds have grown, disbudding can be used to increase bulbil yield. By pinching out the flower buds of certain lily species that do not naturally form bulbils, they can be induced to do so, e.g. *Lilium candidum, L. maculatum*

11. **Crown:** This is a short stem with closely set, short leaves that extends from the central axis above the fruit. These can be cut and planted to produce a single plant. Pineapple crowns are harvested from the fruit or at the time of harvest for propagation. Crowns will bear fruit in around 22 months, while slips will bear fruit in 12-18 months. e.g. Pineapple.

Exercise No. 10

OBJECTIVE - Propagation by Seed in Fruit Plants

SEED TYPES

Seed viability knowledge must be there. Some seed are Orthodox and some are recalcitrant.

1. **Orthodox seeds** are those seeds that could be dried to low moisture content and tolerate freezing temperatures eg Guava, Sapota, Apple, Cherry etc.
2. **Recalcitrant seeds** could not dried below a relatively critical moisture contentand could not tolerate freezing temperature. Desiccation sensitive *e.g.* Jamun, Jackfruit, Mango, Litchi, Mangosteen, Durian Avocado, Citrus, Rabutan etc.
3. **Intermediate seeds:** Which exhibit the drying tolerance characteristic of the orthodox seeds but are sensitive to low temperature storage like recalcitrant seeds *e.g.* Papaya, Macadamia nut.

Propagation of Papaya is Through Seed

- Its seed exhibit dormancy upto 35 days.
- Nursery of papaya is laid during July because proper growth seedlings require humidity in atmosphere.
- Seed are placed shallow otherwise germination is not proper.
- Seed germinate and are having two to four leaves we transplant in small polybags.

Propagation of Jamun by Seed

- Seeds are sown after just harvest in month of July.
- Seedlings are ready for transplanting within a month.

- Since jamun have poyembryony therefore during transplanting from one seed more than one seedling arises.
- Transplant in polybags.

Propagation of Citrus by Seed

- Seeds are sown just after extraction either in rainy season or February month.
- Ready for transplanting after two month.

Exercise No. 11

OBJECTIVE - Study on Propagation by Cutting

Cutting- It is a detached method of propagation in which any vegetative part of the plant is separated and planted to regenerate the missing parts and develop itself into a new plant. This method is commonly used in plants which root easily and readily, thus, multiplication of plant is very quick and cheap.

Purpose- Multiplication of plants by cuttings includes stem, root and leaf cuttings. The stem cuttings are of four types i.e., hard, semi hard, soft wood and herbaceous cutting. The success in propagation by cutting depends upon factors such as conditions of mother plant, parts of the tree where cuttings are made, time of year, care while planting and after care.

Materials required- Secateur, rooting media, nursery bed/pots, khurpi.

A. Stem Cutting

Next to seed, the stem cuttings are the most convenient and popular method of plant propagation. A stem cutting is any cutting taken from the main shoot of a plant or any side shoot growing from the same plant or stem. It is essential for the cuttings to have a sufficient reserve food to keep tissue alive until root and shoot are produced. The shoots with high carbohydrates content roots better. Cuttings from new shoots (less than one year age) root better as compare to older shoot of the plant. Based on maturity of shoots, the stem cuttings are classified as-

1. **Hardwood cutting-** Hard wood cuttings are made from the mature and lignified stems of shrubs and trees.

Procedure- Select one year old shoots current year or of previous season's growth about lead pencil thickness from healthy, vigorous and young plants. The length of cuttings varies from 10-25 cm in length depending upon species. Each cutting should have at least two or three buds. While preparing the cutting, a straight cut is given at the base of shoot about 0.3 cm below the node while a slanting cut 1-2 cm above the bud is given at the top. Remove the leaves from the cuttings. Treating the cutting with 100-5000 ppm IBA before planting gives better results Make holes in the rooting media/ nursery bed and burry the 2/3 basal portion of cutting in the holes at 45 degree angle facing slant portion to sun in the east. Press the soil around cutting firmly. Sprinkle water as and when necessary. Record the data as per technical programme. It is commonly practiced in Grape, fig, pomegranate, mulberry and phalsa.

2. **Semi hardwood cutting-**Semi hardwood cuttings are prepared from semi matured 6-9 month old, slightly woody shoot. These are succulent and tender in nature and are usually prepared from growing wood of current season's growth. The length of cutting varies from 10-20 cm. The cuttings are prepared by trimming the cutting with straight cut below a node. However, it is better to retain two to four leaves on the top of cuttings. Treating the cutting with IBA before planting gives better results in guava, lemon etc.
3. **Soft wood/ Green wood cutting-** Softwood cutting is prepared from soft, succulent and non-lignified 3-6 month old shoots which have not become hard or woody. Usually the cutting size is 5-15 cm but it varies from species to species. Usually few leaves are retained and before planting, treatment with auxin (IBA) is beneficial. This is commonly used for root stocks of apple, peach, plum and cherry in mist condition.
4. **Herbaceous cutting-** The cuttings are prepared from terminal soft, succulent and tender portion of 1-3 month old shoots of current growth under mist condition ensuring warm and humid condition. This is commonly used in ornamental plants.

B. Leaf/Leaf Bud Cutting

Leaf cutting should preferably be prepared during growing season because buds if inter in dormancy may be difficult to force to active stage. A leaf bud cutting consists of a leaf blade, petiole and shoot piece of stem with attached axillary bud of active growing leaves. In this cutting, 1-1.5 cm stem portion is used when propagating material is small. Leaf bud cutting are best made from material having well developed bud and healthy actively growing leaves. High humid condition is essential for better success in leaf cutting.eg. black berry, lemon, rasp berry.

C. Root Cutting

This is commonly used in apple, pear, cherry, guava, black berry, fig, rasp berry, wood apple etc. The root cuttings should be taken from root pieces of young stock plants in late winter or early spring when the roots are well supplied with stored foods but before new growth starts. It is important with root cuttings to maintain correct polarity when planting as the new shoots develop from the proximal end ie. from the part close to the crown. The proximal end of the root piece should always be up.

D. Others

Some different kind of cuttings are also used by propagators like Basal cuttings, Heel cuttings, Bud cuttings, Eye cuttings, Inter nodal and nodal cuttings, Irishmam's cuttings and Piping cuttings.

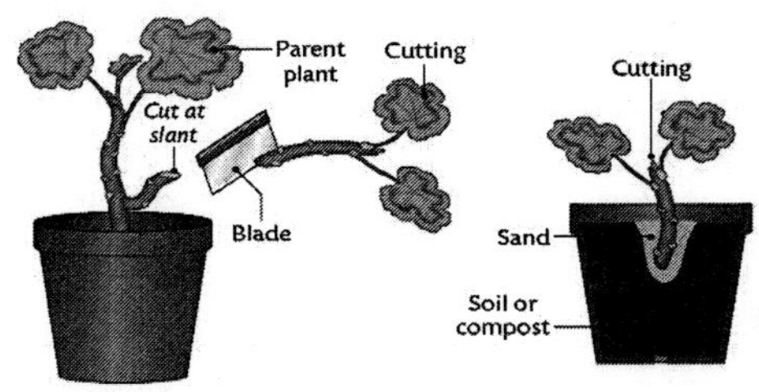

Herbaceous Cutting

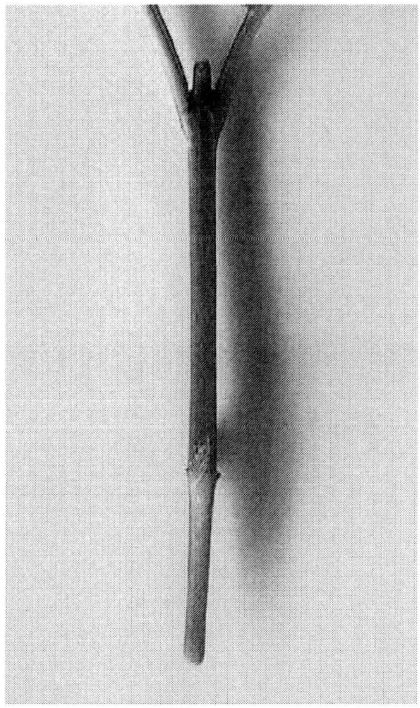

Soft Wood Cutting

Semi Hard Wood Cutting

Hard Wood Cutting

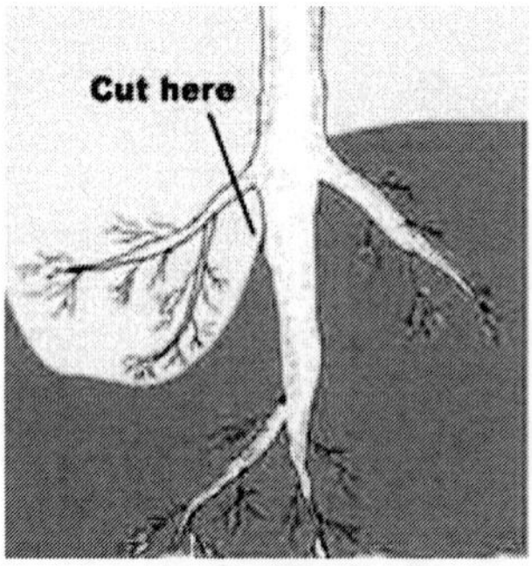		
Root Cutting	***Leaf Cutting with petiole***	***Leaf cutting without petiole***

Exercise No. 12

OBJECTIVE - Study on Propagation by Layering

Layering- The layering is the development of roots on a stem, while it is still attached to the parent plant. The rooted stem is then detached to become a new plant growing independently on its roots.

Purpose- Layering is the oldest technique used by nurserymen to propagate many horticultural plants. Plant multiplication through layering includes several forms of ground and aerial layering (Gootie).When branches running parallel to the ground are utilized, then the method is known as ground layering. When rooting is encouraged on the aerial part of the plant after girdling, then the method is called as air layering or gootie or marcottage.

CLASSIFICATION OF LAYERING

A. Ground layering

1. Tip layering
2. Simple layering
3. Compound or serpentine layering
4. Trench layering
5. Mound or Serpentine layering

B. Air layering (Gootie or Marcottage)

Materials required- Secateur, budding knife, rooting media, nursery bed/pots, khurpi, sphagnum mass, polythene strip, sutali etc.

Propagation by Layering

1. Tip Layering

In tip layering, the tip of shoots is bend to the ground and the rooting takes place near the tip of current season's shoot. The tips of shoot buried 5-6 cm deep in the soil. Keep the soil wet where cane is buried for developing the roots. Rooting in the buried shoots takes place within a month. The new plants may be detached and transplanted in the nursery during spring. eg. Black berry, raspberry and gooseberry.

2. Simple Layering

In simple layering, the flexible shoots of a plant are bent downwards over to ground in early spring or in rainy season. Remove a ring of bark or make a notch at a distance of 20-25 cm away from the tip to encourage rooting. The girdled portion is buried up to 7-10cm depth and covered with soil leaving the terminal end of the branch open. It is necessary to hold the cane/shoot in place with wire or wood stakes. Keep the soil wet where cane is buried for developing the roots. Rooting in the buried shoots takes place within a month. Eg. Grape, lemon etc.

3.Trench Layering

In this method it is important to establish a permanent row of plants to be propagated. The mother plants are planted at the base of a trench at a angle of 45" in rows. The long and flexible stems of these plants are pegged down on the ground to form a continuous line of layered plants. The young shoots that arise from these plants are gradually mounded up to a depth of 15-20 cm in autumn, winter or end of the season, depending upon the species. eg Apple rootstocks (M16 and M25), cherry, plum.

4. Compound or Serpentine Layering

It is suitable for plant producing long, slender, and flexible shoots. It is modification of simple layering in which one year old branch is alternatively covered and exposed along its length. The stem is girdled at different point in the underground. However, the exposed portion of the stem should have at least one bud to develop a new shoot. After rooting, the section are cut and planted in the field eg. Jasmine, American grapes etc.

5. Air Layering

Generally one to two years old, healthy and vigorous shoots having pencil thickness are used for air layering. First the leaves are removed near the basal – inter nodal portion which is away from 35 to 45 cm from apex of the selected shoots then the stem is given a notch or is girdled by removing a ring of bark about 2-3 cm wide. Application of root promoting hormones at the distal end at time of layering helps to get profuse rooting within a short time. Root promoting substances may be applied as powder or in lanolin or as a solution. IBA or a combination of IBA + NAA, both at the rate of 500ppm may be applied for better results. After application of hormones, ringed or girdled portion is covered with moist moss grass or handful of moist clay soil. This ball of earth may be again covered with sphagnum moss and wrapped with a 200 gauge polythene sheet. Air layering should be done either in spring or in monsoon. The rooted layers are either planted in pots or in the nursery beds in a shady place until they are fully established. Litchi, guava and pomegranate, lemon and Lime can be propagated by air layering.

6. Stooling/Mound Layering

In this method the mother plants are headed back to 10-15 cm above ground level during dormant season. The new sprout will arise within two months. These sprouts are then girdled and rooting hormone made in lanolin paste is applied to the upper portion of the ring. The concentration of rooting hormones are varies from plant to plant but in general 3000 to 5000 ppm is most commonly used. These shoots are left for two days for proper absorption of hormone before they are covered with soil. Care must be taken to keep the soil moist all times. The roots from shoots may emerge within 20-30 days depending on species. These rooted stools should be separated from the mother plant only after 60-70 days and then planted in the nursery beds. Eg. Guava, Apple rootstocks, quince, currants, raspberry etc.

Types of Layering

Air layering

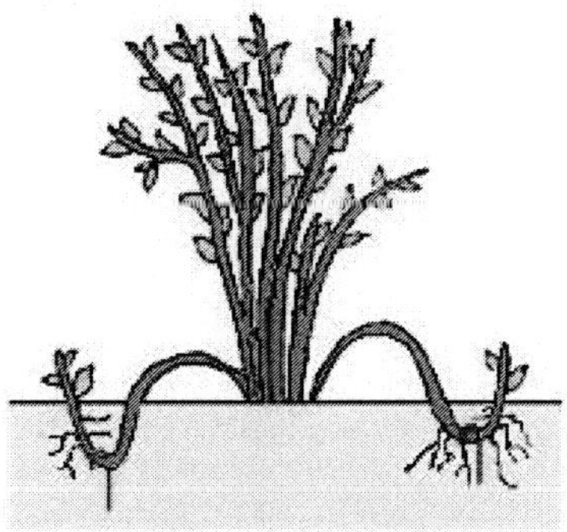

Simple layering

Tip layering

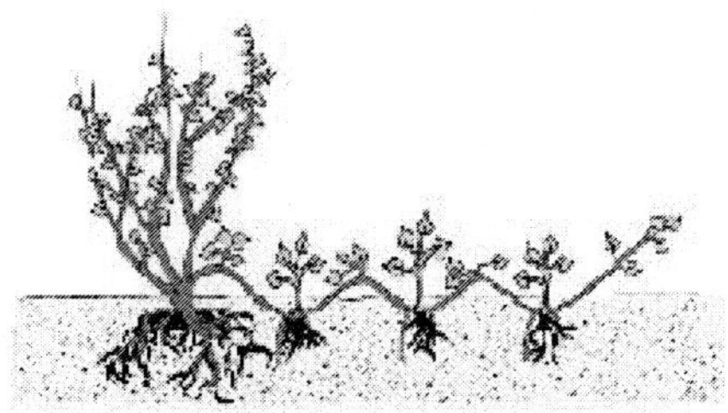

Compound & Serpentine layering

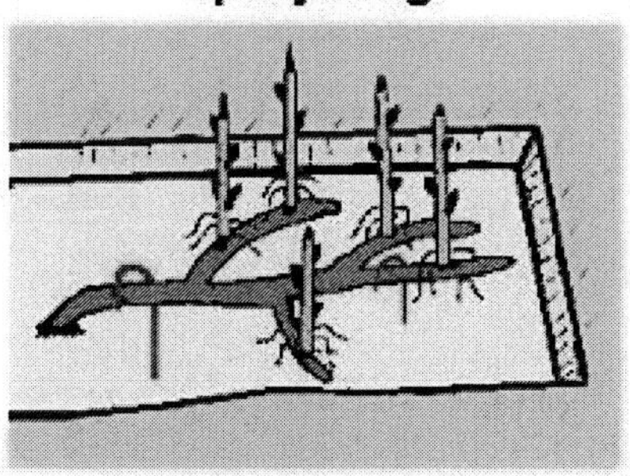

Trench layering

Mound or Stool layering

Exercise No. 13

OBJECTIVE - Study on Propagation by Grafting

GRAFTING

Many horticultural plants are propagated by grafting. In grafting, the desired cultivar can be raised on other plants (rootstocks) for achieving the desired benefits.

Grafting is an art of joining the stock and scion in close contact with each other in such a way that they will unite and continue to grow as single individual/composite plant. The upper part of the composite plant is termed as 'scion' and the part which forms the root is termed as 'rootstock'. Sometimes, when scion and rootstocks are not compatible with each other, another piece of wood is used in between the stock and scion, which is compatible with both; this is called as 'interstock'.

Principles of Grafting- The principal steps involved in healing process and formation of the graft union are-

i. Establishment of direct contact between the cambial region of both stock and scion.

ii. Production and interlocking of parenchymatous cells.

iii. Production of new cambial cell.

iv. Formation of new vascular tissues.

DIFFERENT METHODS OF GRAFTING

A. Attached method of grafting- Inarching, bridge grafting

B. Detached method of grafting – Whip, tongue, cleft, veneer, soft wood, ep-icotyl grafting.

1. Inarching

It is generally used for repairing or replacing damaged root system and hence also called as repair grafting. Selection of parent tree for taking the scion is an important factor for its success. The scion plant should be healthy, vigorous and high yielding.

The stock is brought close to the scion. A thin slice of bark (6-8 cm long and about 1/3 inch in thickness at height) at about 20 cm above the ground level is removed from the stock with a sharp knife. A similar cut is made in the scion. Thus the cambium layers of both stock and scion are exposed. These cuts are brought together and tied firmly with the help of polythene strip. After successful union, stock above and scion below the graft union are looped of gradually. It is done soon after rainy season provided that temperature of the localities does not fall below the 15 0C. eg Mango, sapota, guava,litchi.

2. Veneer Grafting

It is simple method of propagation and can be used in one year old rootstock seedlings having a diameter of 1.0-1.5 cm. For veneer grafting, 3-6 months old scion shoots are selected. Usually, the terminal and next to terminal shoots are most ideal. The shoots are defoliated 5-10 days prior the grafting leaving the petiole attached. The rootstock is prepared by making a slating cut (5cm long) and an oblique cut is made at the base of first cut so that a piece of wood along with bark is removed. The base of the scion wood is then fitted into the rootstock in such a manner that the cut surface including the cambium layers of scion and rootstock face each other. The rootstock and scion are tied together with polythene tape. When scion growth begins the shoot of rootstock is removed above the graft union. Eg. Mango.

There are some other methods of grafting used for propagation of fruit plants.

3. Whip Grafting

It is simple and popular method of grafting. In this method of grafting, it is essential that both stock and scion should be of equal diameter1-1.5 cm. About one year old rootstock is headed back at a height of 20-25 cm from the soil and a diagonal cut is made at the distal end of the rootstock. A similar slanting cut of 2-4 cm is made on the proximal end of the scion. The cut surface of both rootstock and scion are bound together and tied firmly with polythene tape or banana fibre. Many fruit plant are propagated by whip grafting eg.apple and pear etc.

4. Tongue Grafting

This method is practiced as whip grafting except one additional reverse cut is made on both scion and rootstock, so that cambial contact will be more with more success percentage.

5. Cleft Grafting

It is particularly suitable in rootstock having diameter greater than the scion. Rootstock with 5-7 cm or more girth is selected for this purpose. The rootstock is cleft grafted after decapitating the stock 20-40 cm above the ground level. The beheaded rootstock is split to about 5cm deep through the center of stem. After that a hard wooden wedge is inserted to keep open for the subsequent insertion of scion. The scion of 15-20 cm size is taken from a terminal shoot, which is more than three month old and then it is wedge securely (6-7 cm). The cleft of the scion then slipped into the split of the stock. In thicker rootstock more than one scion should be inserted. The graft should be thoroughly waxed to prevent wilting. eg Avocado, apple, pear, plum, mango.

6. Bark Grafting

It should be done in spring when bark of the stock slips easily. It is important that scion used in bark grafting should be dormant. The stock is first sawed off at a point, where bark is smooth. Bark is split downward, about 5 cm from the top. Scion of 10-12 cm long, containing 2- 3 buds are collected from the dormant wood and are preparing by giving slating cut (5cm) downward along one side of the base. The prepared scion then inserted in the center of split between the bark and wood of the stock. The scion is kept firmly by using adhesive tape. eg Many fruit plant.

SPECIAL GRAFTING TECHNIQUES

1. Epicotyl /Stone Grafting

This method is commonly practiced in mango. It is simple method of wedge grafting where the current season's scion shoot of the desired variety is inserted into the tender part of the stem of the sprouted stone (15-30 days old seedling) and tied with polythene sheet

2. Top Working

This is commonly practiced to convert an old/ unproductive orchard of inferior variety in to productive one by grafting with desirable variety after head back of unproductive plant eg. Mango, ber, cashew nut, mulberry

3. Soft Wood Grafting

This technique is commercially used for raising Cashew nut, Mango, Jamun, Tamarind, Custard apple through wedge grafting. In this technique, grafting is done with mature, procured scion on the emerging soft, coppery red shoot of the rootstock, which is 60-70 days old.

4. Bridge Grafting

Bridge grafting is done with objective of repairing of damaged fruit plant. The scions are prepared by giving slanting cuts on one side of the top and base. These scions are inserted above and below the injury of the plant and tied properly.

5. Double Working

Double working is a specialized technique of grafting in which the composite plant has three different components, the root stock, interstock and the scion i.e. the desired variety or cultivar. Thus the double worked plants have graft joints, one between the rootstock and interstock and other between the interstock and scion. It is done to overcome the incompatibility between the desired cultivar and stock.

6. Micrografting

The grafting of tiny plant parts under asceptic and controlled environmental condition is called micrografting. Micrografting has been mostly used in citrus, apple and plum to produce virus free plants.

Inarching

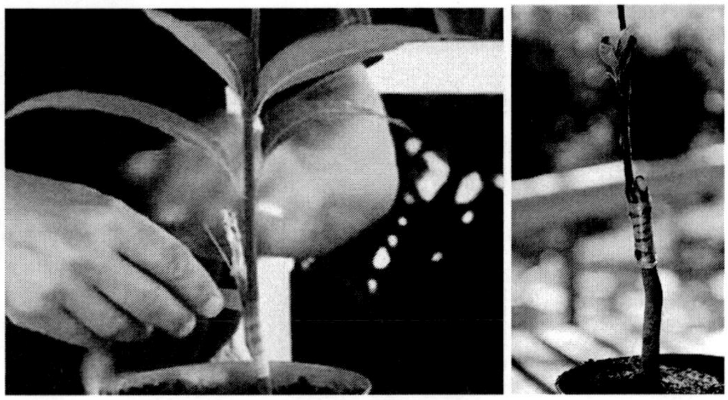

Veneer Grafting

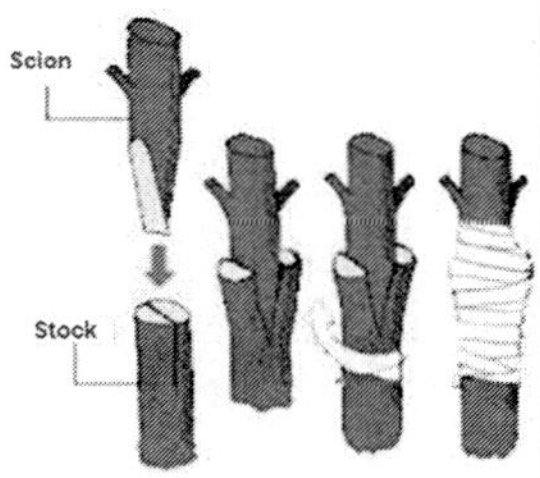

Cleft/ Wedge Grafting

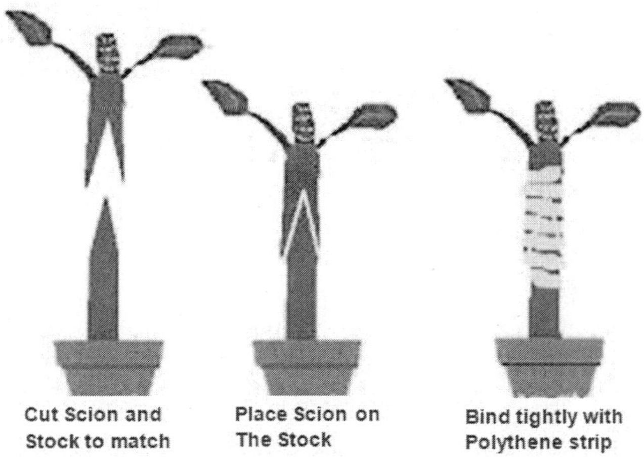

Saddle Grafting Inarching

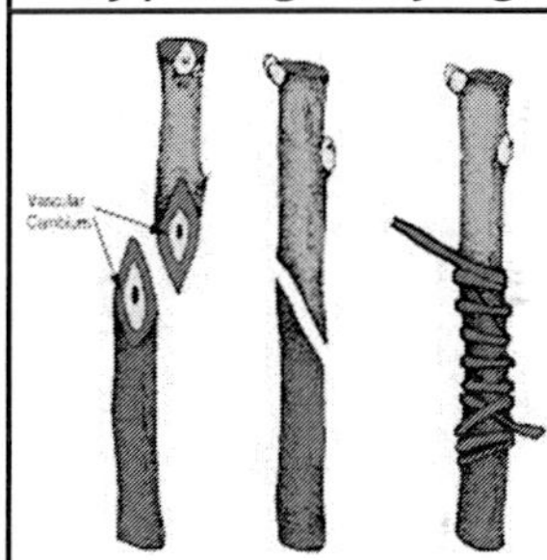

Whip Grafting

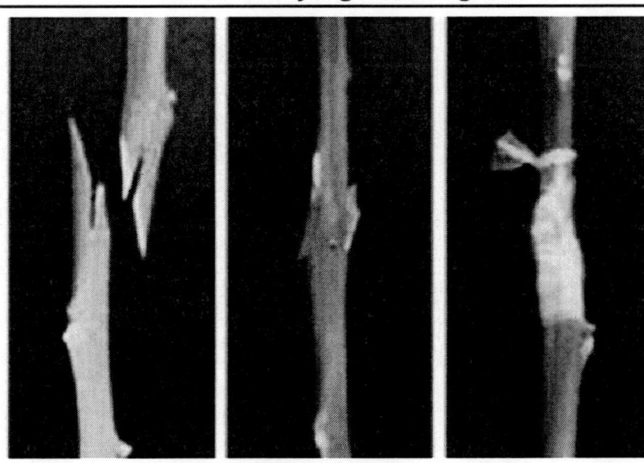

Tongue Grafting

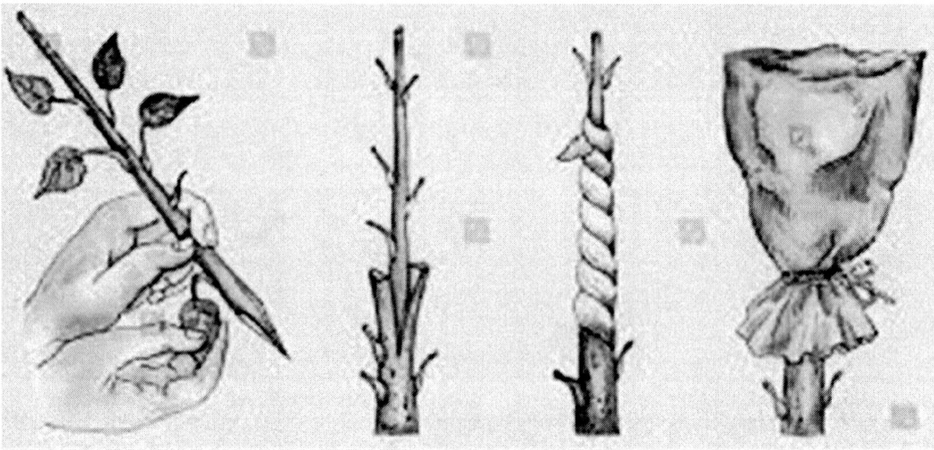

Epicotyl Grafting

Bridge grafting to repair damage

Exercise No. 14

OBJECTIVE - Study on Propagation by Budding

Budding- Budding is also a method of grafting, wherein only a single bud with a piece of bark with or without wood is used as a scion material, which develops into plant after successful union of the stock and bud. Budding is generally done when the stock plant is in active growth and more cambial activity.

Bud Union- Like graft union, a series of changes takes place in the formation of successful bud union also. In general, four stages viz. pre- callus, callus, formation of cambial bridge and healing process etc. takes place for the formation of successful bud union.

Nurserymen employ various method of budding but according to convenience in performing the operations and percentage of success, the following methods are the commercially advocated in propagating various horticultural crops.

1. Shield or T-Budding

A 'T' shaped cut is made on the selected portion of the stock with the help of sharp budding knife on one–year-old rootstock seedling having 2-2.5 cm thickness at 15-20 cm height. The bark of seedling should slip easily. The two flaps of bark are then loosened slightly with the help of budding knife. From the bud wood, which is selected from a healthy shoot of a current season's growth, the buds of middle portion are selected. These are removed from the bud wood by cutting shallowly about 5-6 mm below and 2-3 cm above the bud. This shield piece containing a bud is inserted carefully in 'T' shaped incision made on the stock. This bud then presses firmly and tied with polythene strip. After the bud has sprouted, the stock is cut to about 10-15 cm above the bud. eg citrus, aonla, custard apple, jamun, bael , plum, peach, cherry, ber, rose etc.

2. Inverted 'T' Budding

As the name indicates, the cut is to be given on the root stock is reverse to that of 'T' i.e. inverted 'T' cut is given on the stock. This is widely used in high rainfall areas.

3. Patch Budding

A rectangular patch of bark is removed completely from the one year old seedling stock and replace with a similar patch of bark containing a bud of desired variety. It is successfully used in species having thick bark such as aonla, bael, jamun, guava, walnut, pecan nut.

4. Ring Budding

In ring budding, a complete ring of bark is removed from the stock and it is completely girdled. A similar ring of bark containing a bud is removed from the bud stick and is inserted on to the rootstock. In this budding both scion and stock should be of same size. It is utilized in peach, plum, ber, mulberry etc.

5. Modified Ring Budding

In modified ring budding, complete ring of bark from the both scion and stock is removed by making one vertical cut on the opposite side of the bud, so that bud can be removed easily, eg. aonla, ber etc.

6. Chip Budding

Chip budding is successful method of budding when the bark of the stock does not slip easily. A chip of bark and wood is removed from the smooth surface between the nodes of the stock. A chip of similar shape and size is then removed from the bud wood of desired cultivar. For which, a 2-3 cm long down ward cut is made through the bark and slightly in to the wood of the stock. Then a second cut of about 2.5 cm is made so that it bisects the first cut at an angle of 30-45 0.in this way the chip of wood is removed from the stock. The bud chip then slipped in the place of rootstock from where chip has been removed. eg grape.

7. Flute Budding

In flute budding a patch of bark (Flute) encircling the stock is removed leaving a narrow strip thereon. A similar patch of bark containing the bud is taken from the scion plant and placed on the cut surface of the rootstock followed by tying as usual when the bud exhibits the signs of growth, the top of the stock is cut back.

8. Forkert Method of Budding

It is modified method of patch budding, in which the bark flap of the patch is not removed from the stock plant but used to cover the bud on the stock plant. The bud is inserted in to the flap. It is covered with flap of bark on the stock plant and tied firmly with alkathene strip. Example- Aonla, ber bael and guava.

Types of Budding

Types Used When Bark is Slipping

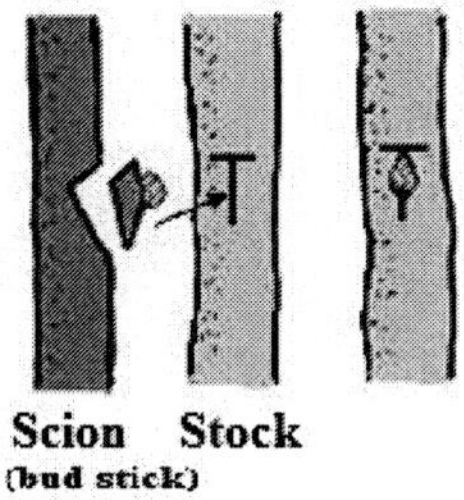

T-budding

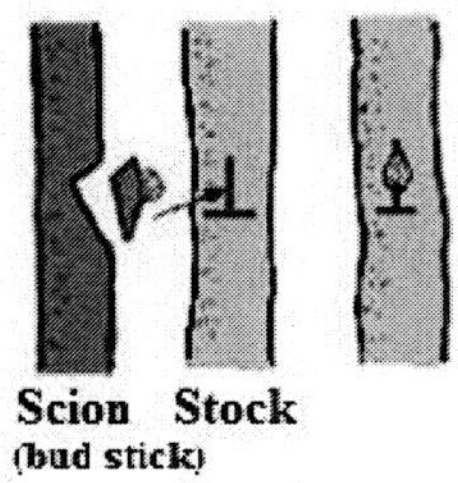

Inverted T-budding

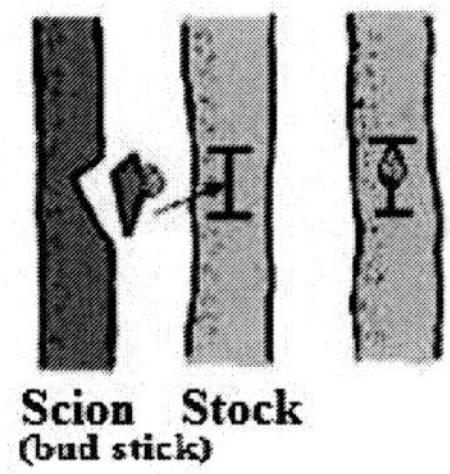

I-budding

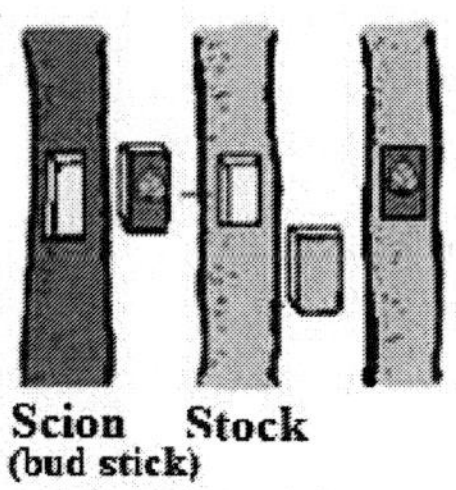

Patch-budding

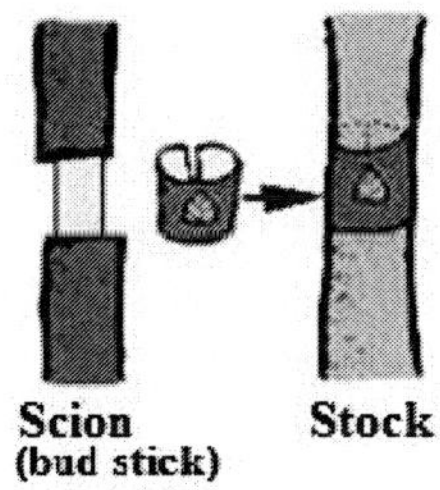

Ring budding

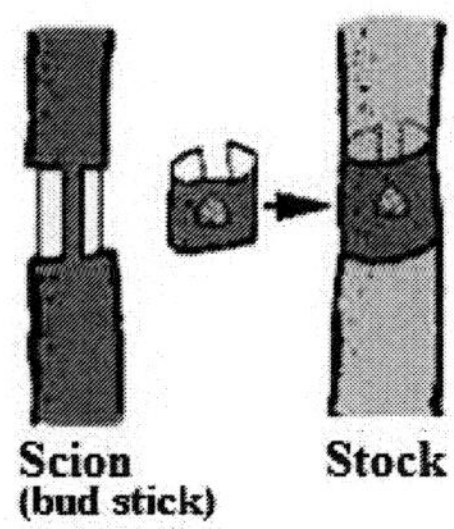

Flute budding

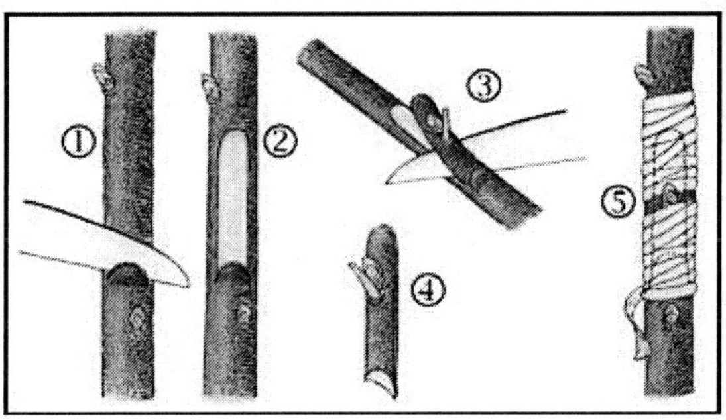

Clip Budding

Exercise No. 15

OBJECTIVE - Study about Training and Pruning

Training and pruning are an important activity in fruit crops to have better frame work and optimum fruiting area. Training refers to giving a desired shape to the plants by tying or staking or supporting over a structure and or selective pruning for a good strong frame work. Pruning refers to cutting of certain portion of plants for maintenance of fruitfulness and quality besides vigour of the trees or vines. Pruning affects the functions of the plants and assists in better fruiting and in getting more quality fruits. It is one of the most crucial operations and requires some scientific knowledge regarding bearing behavior of the plants.

OBJECTIVES OF TRAINING

The major objectives of training are

i. To give a strong frame work to the tree for supporting good cropping.

ii. Provide good exposure of light and air to branches and leaves.

iii. To maintain tree growth in such a way that that various cultural operations, such as spraying, annual pruning, harvesting etc can be done at the lowest cost.

iv. To protect the tree from sun burn and damage.

v. To secure a balanced distribution of fruit bearing parts on them in limbs of the tree.

vi. Maintain the vitality of trees over a long period of time.

Principles of Training

- Training should be started from very beginning age of the plant.
- Most of the trees are trained in single stem system. While, in some fruit crops like anar, fig, custard apple and bushy plants are trained through multi stemmed training system.
- Through training can control apical dominancy with balance retain desirable branches on tree.

- ◈ The shoots having narrow crotch angle (<40-45^0) should be discarded.
- ◈ Water sprouts should be removed time to time.

METHODS OF TRAINING

There are three most commonly used training methods are followed in fruit crops based on the growth habit of the fruit tree. These are

1. Open Centre

In this system of training, the main stem is allowed to grow only upto a certain height, thereafter it (leader or main stem) is headed back to encourage lateral branching (scaffold branches). This system is also known as **Vase-shaped system**. This system allows better distribution of sunshine and to reach it to branches of trees and also facilitate cultural operations like spraying, thinning, harvesting etc.

2. Central Leader

In this system of training, main stem (leader) is not headed back and is allowed to grow I its natural ways extending from surface level to the top of the tree. This results in robust close centre and tall tree and branches are more fruitful near the top as compared to lower branches. This system of training is also known as closed centere done.

3. Modified leader

It is intermediate between the open centre and central leader training system. In this system main stem is allowed to grow unhampered for the first four or five years, thereafter it is headed back and lateral branches are allowed to grow as in the open centre system. Modified leader system produces fairly strong and moderately spreading trees.

Trees are trained to different forms with or without the support of certain structures.

Espallier system/cordon system

Telephone system

Open Center System	**Modified leader System**	**Central Leader**

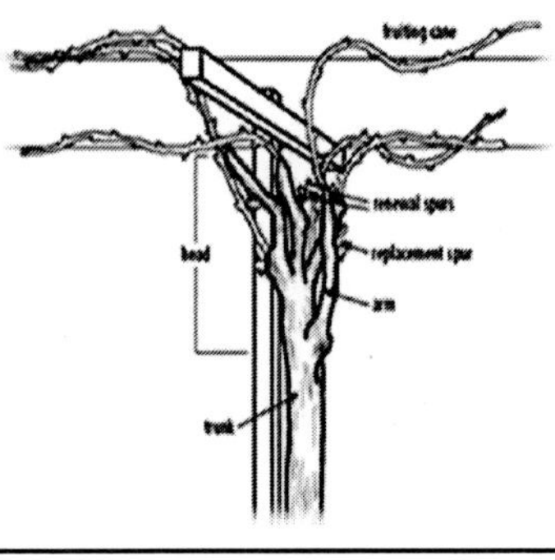	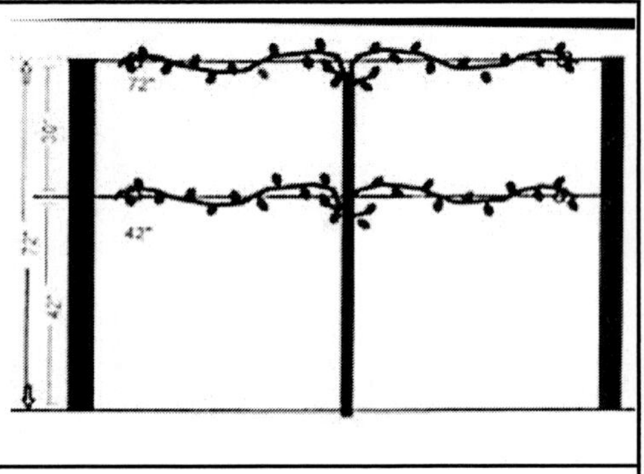
Head system	**Kniffin System**

Bower system

SPECIAL METHODS OF TRAINING

Bower System

It is also called as 'Pandal' or 'Pergola' system. It is generally practiced in grapes and other cucurbitaceous vegetables like snake gourd, ribbed gourd, bitter gourd etc. In this system, the vines are spread over a criss cross net work of wires, usually at 2.1 to 2.4m above ground, supported by concrete or stone pillars or live support like *Commiphera sp*. The vine is allowed to grow single shoot till it reaches the wire net and is usually supported by bamboo sticks tied with jute thread. When the vine reaches the wires, its growing point is pinched off to facilitate the production of side shoots.

Kniffin System

In this system, two trellis of wire are strongly supported by vertical posts. The vines such as grape when trained in this system has four canes one along each wire and the bearing shoot hangs freely with no tying being necessary.

Telephone System

This system consists of 3 or 4 wires usually kept at 45-60 cm apart fixed to the cross-angle arms supported by vertical pillars or posts. Vines are allowed to grow up to a height of 1.5 to 2.0 m and then trained on this system. Moderately vigorous cultivars with apical dominance are best trained on such system.

Head System

It is mostly used for spur bearing grape cultivars. In this system, vines are trained like a small bush. Vines are allowed to, grow up to 1.2 meters, and then headed back to produce laterals. Four laterals- one in each direction is allowed to grow and rest are thinned out. In next dormant season, these laterals are cut back to 2 buds and further two arms of 20-30 cm are allowed on each secondary arm. After 3-4 years these vines will give a dwarf bush like appearance and requires no staking.

Other training systems which require no staking are Palmette, Spindle bush, Dwarf pyramid and Head and spread systems.

Cordon and Espalier System

Plants are trained to grow flat on trellis or on horizontal wires by training the branches perpendicularly to the main stem on both the sides, and trained horizontally on the wires. Plants trained in this system are called **'espaliers'**. An espalier with one shoot or two shoots growing in opposite or parallel directions are called a **'cordon'**.

Tatura Trellis

In this system, trees are trained to a multi-layered wire trellis. The trellis is V-shaped, supported by two long, stout poles embedded into the soil angles of 60^0 from the horizontal. Five wires at 60cm intervals are fastened to these poles. This system is being now followed for pome fruits, nut fruits and grapes. The trees are grown as double leader. Trees with each leader inclined at an angle of 60^0 from the horizontal.

Pruning

Commonly, trees are pruned annually in two ways. A few shoots or branches that are considered undesirable are removed entirely without leaving any stub. This operation is known as 'thinning out'. The other method which involves removal of terminal portion of the shoots, branches or limb, leaving its basal portion intact, is called 'heading back'. Thinning out involving large limbs as in old and diseased trees is called 'bulk pruning'.

Pruning is done with the following specific objectives.

i. To remove surplus branches,

ii. To open the trees so that the fruits will colour more satisfactorily

iii. To train it to some desired form

iv. To remove the dead and diseased limbs

v. To remove the water sprouts and

vi. To improve fruiting wood and to regulate production of floral buds.

Principles of Pruning

- Remove water sprouts in later part of plant life.
- Avoid bark injury while pruning
- Pruning should be completed well in advance of flowering season.
- In deciduous plants, pruning should be done in advance of winter so thatlow temperature injury may be minimized.
- Apply Bordeaux mixture or paste or COC just after pruning to avoidincidence of diseases spread.
- Overcrowded, diseased, damaged and insect infested shoots should be removed.

Season of Pruning

The pome fruits such as apple, plum, pears and peaches are pruned every year in December - January. Under North Indian conditions, the grapes are pruned in first fortnight of January. Under South Indian conditions, old non bearing mango trees are pruned during August – September. Jasmines are pruned to 45cm height from the ground level during the last week of November.

Method of Pruning

- **Heading back**: removal of portion of shoot or branch at a certain portion eg 50 % or½.
- **Thinning out**: complete removal of shoot or branch at its origin or emerging points.
- **Ringing or girdling**: in this process a circular ring of bark measuring about5 mm to 3 cm in length is moved for hastening bearing by allowing greater accumulation of photosynthesis in upward of the plant.
- **Notching**: making a notch above a bud by removing a wedge shaped piece of bark is termed as notching. It checks the influence of hormone and encourages growth.
- **Nicking**: making a notch below a bud by removing a wedge shaped pieceof bark is termed as nicking. This ensuring accumulation of carbohydrates from the leaves to the bud and may result in the formulation of fruit bud.
- **Root pruning**: a portion of coiled, fibrous, old roots are removed from citrus, temperate fruit crops for regulating the flowering.
- **Leaf pruning**: a portion of leaves are pruned to regulate flowering inguava, bonsai for size control and maintain fruit: leaf ratio.
- **Bending**: Erect growing varieties of guava by this technique enhance quality crop.
- **Coppicing:** Practice when removal of main stem near to ground levelleaving 30-35 cm stump alone (cinchona, jamun). The coppiced stump start producing many vigorous shoots in about 6 months time, after that retained 2-3 shoots per stump and rest completely tinned out.
- **Pollarding:** Practice of removing the main stem at certain height (1-1.5 m) to encourage branches above that height.
- **Lopping:** This practice of removing the canopy cover during dormancy or open the centre to permit light etc.
- **Pinching:** Practice to reduce the plant height and to promote auxillary branching.
- **Disbudding:** The practice of removing unwanted flower buds in cluster so as to encourage the remaining buds to develop into a large, showy, quality bloom is called disbudding.

- **Smudging:** It refers to the practice of smoking the tree like mango, commonly employed in Philippines to produce off season crop. Smudging of mango trees in India has not been found to induce early blossom.

Thinning

Fruit thinning is an exhaustive process to the tree especially if the crop is heavy. The other objectives of fruit thinning are the following:

1. To increase the annual yield of marketable fruit.
2. To improve the fruit size.
3. To improve the colour of the fruit.
4. To improve the quality of fruit(T.S.S.)
5. It reduces the limb breakage.
6. It promotes tree vigor and ensures more regular cropping.
7. It permits more thorough spraying and dusting of fruits during the late season application.
8. It ensures uniform ripening.

Methods of Thinning

1. Hand thinning
2. Chemical thinning
3. NAA at 100 ppm reduces the fruit setting from 67% to 50% in Anab-e-Shahi variety of Grapes.
4. In mandarin, NAA 600 ppm on marble sized stage is recommended to thin the over bearing fruits so as to increase the size and quality of fruit.

Time of thinning- At blossom time, at pea and marble stage of fruit, soon after the natural fruit drop of young fruits has started.

Pruning of Ber, Grape, Phalsa, Fig, Pear, Peach:

Crop	Time	Level	Remarks
Ber	15 April – 15 May	Sixth secondary branch level	Spraying of 3% thiourea or potassium nitrate once in 2 days before pruning induces bud sprouting from maximum number of nodes.
Grape	Winter (Dec-Jan)	1-2 buds in renewed and 12 buds are retained on fruiting canes	The fruited canes are pruned to renewal spurs and pruned to fruiting canes in the nextwinter.

Crop	Time	Level	Remarks
Phalsa	January	50-100 cm from the ground level	Hard pruning in North and light pruning in South India require. Pruned phalsa canes can be utilized for making of baskets to transport fruits and vegetables and as stakes for support of vegetables or fruitplants.
Fig	Winter (4-5month advance)	Light pruning	After fruit harvesting light head back or Notching on 8 month old can at middle part 2.5 cm long cuts. Heavy pruning where trees are headed back severely every year, leaving about 2 buds on each one-year- old shoot.
Peach	Dormantpruning	20-30% linear growth annually	Bearing laterally so by thinning out and heading back of the shoots achieve good canopy.
Pear	Dormant	30-50% level	Spur bear crop that is 2 yearold branch so balance between fruit production and vegetative growth.

Exercise No. 16

OBJECTIVE – Study of Manures and Fertilizers Application

In a general way, manures and fertilizers are the substance containing plant nutrients, which is applied to the soil for increasing crop production.

MANURES

Manures are plant and animal wastes that are used as source of plant nutrients. They release nutrients after their decomposition. Manures can be grouped into bulky organic manures and concentrated organic manures.

a. Bulky organic manures - Farm Yard Manure (FYM), compost from organic waste, night soil, sludge, sewage, green manures.

b. Concentrated organic manures - oilcakes (edible, non-edible), blood meal, fishmeal and bone meal.

FERTILIZERS

Fertilizers are industrially manufactured chemical containing plant nutrients. Nutrient content is higher in fertilizers than organic manures and nutrients are released almost immediately. The fertilizers have three groups;

Straight fertilizers – supplies single nutrient Ex: Urea, Muriate of Potash
Complex fertilizers - supplies two or more nutrient Ex: 17:17:17 NPK complex
Mixed fertilizers- supplies two or more nutrient Ex: Groundnut mixture.

Bio- Fertilizers

Biofertilizer is an organic product containing a specific microorganism which has ability to convert unavailable nutrients in to available form through biological process eg. *Rhizobium*, *Azotobacter*, *Azospirillum*, Blue Green Algae, PSB,

ROLE OF MANURES AND FERTILIZERS

1. Organic manures bind the sandy soil and improve its water holding capacity.

2. Organic manures open the clayey soil and help in aeration for better root growth.
3. Organic manures add plant nutrients in small percentage and also add micronutrients, which are essential for plant growth.
4. Manures increase the microbial activity which helps in releasing plant nutrients to available form.
5. Organic manures should be incorporated before the sowing or planting because of slow release of nutrients.
6. Fertilizers play an important role in crop production as they supply large quantities of essential nutrient to crops
7. Fertilizers are manufactured in forms that are readily utilized by plants directly or after rapid transformation.
8. Fertilizers dose can be adjusted to suit the requirement as determined by soil testing.
9. Balanced application of nutrient based on crop requirement is possible by appropriate mixing of fertilizers.
10. Fertilizers applied as straight fertilizers (providing single nutrient) or complex and mixed fertilizers (supplies two or more nutrients) based on crop requirement.

Enhancing fertilizer use efficiency-The following are the agronomic measures to improve the Fertilizer use efficiency (FUE).

1. Using best fertilizer source
2. Using adequate rate & diagnostic techniques
3. Use of balanced fertilization
4. Integrated nutrient management
1. Utilization of residual nutrients

NUTRIENT ASSESSMENT IN FRUIT CROPS

Management system adopted for fruit trees are quite different, especially with regard to nutrient than that adopted for field crops. The shift in management system could be commenced 3-4 years after planting depending on the tree size. Nutrients are required for flowering and fruiting while at the same time trees are allowed to grow and maintain sufficient vigour for producing high yields in following years. To maintain productivity of trees in long run and maintain the sustainability of tree production capacity, application of nutrient should be based on actual requirement and availability of nutrient in the soil. Application of plant nutrients economically at correct time with right amounts in a way that nutrients could be taken up by plants efficiently with minimum losses. The purpose of assessment of nutrient requirement of fruit trees is to keep mineral nutrient levels in the tree with in the desired range to have the growth and development effects and fruiting of trees as desired.

The most important thing in nutrient management for bearing trees is to analyze the importance of timing of nutrient application in relation to tree phenology or growth cycle.

Factors Influencing Nutrient Content of the Soil

- Type of vegetation cover; e.g. with legume cover there could be higher N, etc.,
- Application of manures and fertilizers.
- Application of soil amendments.
- Inherent status of soil

Factor affecting Nutrients Content of the Leaf

- Fertilizers input
- Type of vegetation cover; e.g. with legume cover there could be higher N, etc.,
- Factors inherent in the tree; genetic make-up, age of crop, time of sampling; yearly (seasonal) variation, age of leaves, position of leaves, yield of crop.

Leaf Samples

Leaf analysis is an important tool to estimate nutritional requirement of fruit trees. Leaf nutrient content can be obtained by analyzing leaf tissues at proper growth stage. Thereafter, nutrient requirement of fruit trees can be calculated based on optimum norms of tissue nutrient in different fruit crops.

Application of manures and fertilizers: It depends mainly on basic principles of what to apply, how much to apply, how to apply, when to apply and where to apply.

Methods of Application of Solid Manures and Fertilizer

Fertilizer recovery is greatly influenced by method of its application. Method of application varies according to the spacing of crop, type of fertilizer material, time of application. Various methods of application of fertilizers are in vogue and following are few important practices widely adopted.

A. Broadcasting

It refers to spreading fertilizers uniformly all over the field on entire area or in basins. It is of two types-

i. **Basal dressing-** Spreading of fertilizers before sowing or planting of the crops and mixing them by cultivating the soil during seed bed preparation is termed as basal application through broadcasting.

ii. **Top dressing and side dressing-** Spreading of fertilizer in standing crops without considering the crop rows is termed as top dressing. But when the crop rows are taken into account and the material is dropped on the ground surface near the crop rows then it is called as side dressing.

B. Placement

This refers to applying fertilizers into the soil from where the crop roots can take them easily. Placement could be done in following ways:

i. **Plough sole placement**-When the fertilizers are applied in open furrows at plough sole level while ploughing then it is termed as plough sole placement. Such furrows are covered immediately during the next run of the plough.

ii. **Deep placement**-The method is adopted in dry land condition where the fertilizers are placed deeper than plough sole level then it is called as deep placement.

iii. **Sub-soil placement-**When fertilizers are placed still deeper than the seeding or planting depth and also deeper than the previous two methods the method is termed as sub-soil placement.

C. Localized Placement

There is distinction between placement and localized placement. The former refers to applying fertilizer into the soil without special reference to the location of seed or plant while the latter implies the application of fertilizer into the soil close to the seed or plant. The method could be adopted in following ways:

a. **Contact placement/combine drilling-**When fertilizer is placed along with seed then it is called as contact placement. This is done by using seed-cum-fertilizer drill. Sometimes fertilizer is drilled by implement and seed is sown in the same furrow.

b. **Band placement-**This is a localized placement of fertilizers by the side of plants or seeds (about 5 cm apart). This may be of two types as the bands may be continuous or discontinuous:

i. **Hill placement (discontinuous band)-**In the hill for widely spaced plants like cucurbits, and castor etc fertilizers are placed on either of both sides of plants along or across the row but not along the entire row. This method is also termed as discontinuous band application.

ii. **Row placement (continuous band)-**Along the entire rows of closely spaced crops like vegetables-potato and flowers fertilizers are applied continuously at 2-2.5 cm depth. This method has a definite relationship of fertilizers with seedlings or seed as the fertilizer is placed to the side of seedlings or seeds some distance away from them or at the level of the seed, above or below or by the side of the seed level. When the soil surface is dry, this method gives very promising results.

iii. **Ring placement (continuous band)-**Fertilizer is applied in a circle around individual plant or hill base at a depth of about 2.5-5 cm. It is most common method in fruit crops.

D. Pocket/spot Placement

When fertilizers are placed at a fixed spot by the help of a bamboo peg having a hole at the bottom in case of very widely spaced crops then the method is termed as pocket/spot placement method. Fertilizers are placed deeper into the pocket (dibble) and seeds are sown in the same pocket about 5 cm above the fertilizers.

E. Pellet Placement

This method is adopted specially in case of deep water(rice cultivation) where it is difficult to apply fertilizers in normal methods as the fertilizer granules get dissolved in water before reaching to the ground level. In this method fertilizers (especially nitrogenous ones) are mixed with clay soil in the ratio of one part of fertilizer into 10- 15 parts of soil. The fertilizer is well mixed with soil after slight moistening then filled in gunny bags and stored for two-three days. Now small mud bolls are prepared and these boll or pellets are dropped near the crop rows under deep water conditions.

Method of Application of Liquid Fertilizers

Use of liquid fertilizers is not very common practice but in advanced countries, this is the most common method. It is the most suitable method under dry land agriculture and in the areas which are prone to erosion problems. Liquid fertilizers may be applied in following ways:

1. **Starter solution**- As the name indicates, this type of fertilizer application helps the plant in starting the growth quickly after plant establishment. Starter solutions usually contain N, P, K in 1: 2: 1 or 1: 1: 2. This method is used for transplanted crops where in place of irrigation water this solution is applied just to wet the field.
2. **Fertigation**-The required quantity of fertilizer material is dissolved in irrigation water and can be used in surface, sprinkler or drip irrigation systems. This method has got popularity after advent of drip irrigation.
3. **Nutrient injection method** - In USA and some other countries anhydrous ammonia is injected into the soil at a depth of about 20-25 cm and at a pressure of about 200 pound per square inch. Injecting hormonal solution and some micro nutrient solutions in the phloem region of the fruit trees is also becoming a distinct possibility in correcting the nutrient deficiency.
4. **Foliar spray of nutrient solutions**-In this method of fertilizer application urea, micro nutrients and other required materials are dissolved in water, filtered and sprayed over the crop foliage by the help of a suitable sprayer.
5. **Aerial application**-In areas where ground application is not practicable, the fertilizer solutions are applied by aircraft particularly in hilly areas, in forest lands, in grass lands or in sugarcane fields etc.

Time of application: Depending upon crop, annual, biennial or perennial or variety, stage of crop growth etc, the time of application varies. The time of application also varies according to type of nutrient one would like to supply. Usually FYM/compost is applied at the time of land preparation at least 15 days before sowing or planting in respect of annuals or short duration crops. However, in the case of perennials, they are applied to pits at the time of planting and to basins during subsequent years. Similarly, if a green manure crop is grown it will be ploughed back to the soil before flowering and well before the crop is sown/ planted to ensure proper decomposition and availability of nutrients for early growth and development. The P & K are applied basally at the time of sowing or planting. But nitrogen is applied in varying number of splits depending on the crop and its duration. In any case fertilizers should be made available at right time in right quantity.

CALCULATION OF QUANTITY OF FERTILIZERS BASED ON NUTRIENT REQUIREMENT

Quantity of fertilizer = (quantity nutrient required x 100) nutrient present in fertilizer $Q = (N_1 \text{ x}100)\ N_2$

Where,

Q = Quantity of fertilizer

N1 = quantity nutrient required N2 = nutrient present in fertilizer

Or

Q = quantity of nutrient required x factor

Calculation of Quantity of Complex Dertilizers

First find out the quantity of fertilizer required for supplying of whole quantity of major nutrient present in the particular fertilizer. Thereafter, calculate the amount of second nutrient supplied through the calculated quantity of fertilizer. Then subtract calculated amount of nutrient from the whole amount of second nutrient. Calculate the quantity of another source of fertilizers for balance nutrient quantity of second nutrient.

Calculation of Quantity of Chemicals for Spray Solution

$V_1 = (C_2 \text{ x } V_2) \text{ x } C_1$

Where,

V_1 = quantity of chemical or commercial product V_2 = volume of spray solution to be prepared

C_1 = nutrient content in chemical or commercial product C_2 = concentration of spray solution

Or

$$\text{Desired quantity of Chemical} = \frac{\text{Qty. of spray solution required X Concentration of solution}}{\text{\% of actual ingredient of nutrient in chemical}}$$

Factor for Calculating Quantity of Different Fertilizers

Fertilizers	Factor
Nitrogenous fertilizers	
Urea	2.17
Calcium Ammonium Nitrate	3.85
Ammonium Nitrate	2.94
Ammonium Chloride	4.00
Calcium Nitrate	6.45
Ammonium Sulphate	4.84
Phosphatic fertilizers	
Single Super Phosphate	6.25
Rock Phosphate	5.56
Bone Meal	5.00
Potassic fertilizers	
Potassium Sulphate	1.92
Potassium Chloride	1.66
Potassium Magnesium Sulphate	4.55

Exercise No. 17

OBJECTIVE - Preparation of Growth Regulators Solution

Plant growth regulators or plant regulators are the organic chemical compounds whichmodify or regulate physiological processes in an appreciable measure in the plants when used in small concentrations. They are readily absorbed and they move rapidly through the tissues when applied to different parts of the plant.

GROWTH REGULATORS

1. **Auxin:** Auxin like substances are produced in buds, tips of stem, roots etc. Some of the synthetic substances having auxin like activity are IAA, NAA etc. Main action of this auxin like substances is cell, division, cell differentiation etc.
2. **Gibberellins:** This kind of substances stimulates growth in tissues of young internodes. It acts by modifying RNA produced in nucleus i.e. it has control over cell elongation. Cell elongation or by hydrolysis of starch leads to increased concentration of sugar in cell sap and make entry of water and finally it stretches the cell size.
3. **Cytokinins:** This type of chemicals interacts with auxins. It acts on cell initiation/cell division. When cytokinin to auxin ratio is low, root development will be more. If the ratio is more, shoot development will be more. The prevalence of equal ratio leads to undifferentiated callus production.
4. **Ethylene:** Ethylene is the only gaseous hormone which stimulates growth. Itis natural product used in ripening of fruits. Ethylene has simplest chemical structure $CH_2=CH_2$ and synthesized from methionine amino acid.
5. **Growth inhibitors:** these are synthetic compounds which reduce the growth of the plants. Growth retardants treated plants appear normal through their stems are shortened. Example: AMO-1618, phosphon-D, Cycocel, B-995, CCC, B-Nine, MH-40 etc.

Besides above, there are some other compounds such as brassino-steriods, tricontanols, morphactins and several other phenolics which regulate the plant growth and its physiology.

Application of PGRs: These plant growth regulators commercial use to induce rooting of cuttings (IBA), seed dormancy breaker (GA_3), weed control (2, 4- Dichloro phenoxy acetic acid (2,4-D), fruit drop control (NAA), fruit qualityenhance (GA_3) etc.

Some of the commercial uses of plant regulators in fruit crops are as follows:

- To break seed dormancy and enhance and enhance germination: GA_3
- To promote rooting of cutting and layers: IBA and NAA
- To propagate horticultural plants through tissue culture: BAP, Cyto-kinin,GA_3
- To break dormancy: GA_3
- To induce flowering: NAA, Ethrel
- To prevent flower and fruit drop: NAA, 2,4- D
- To overcome alternate bearing: Paclobutrazol (PP_{333})
- To induce parthenocarpy: 2,4-D, NAA
- To improve parthenocarpy: 2,4-D, NAA
- To improve fruit size and fruit quality parameters: GA_3
- To regulate ripening (i.e. delaying or promoting) process: Ethylene
- To control weeds: 2, 4-D, 2,5,4-T

METHOD OF APPLICATION OF PLANT GROWTH REGULATORS

Growth regulators are generally applied at very low concentrations i.e. ppm (parts per million). One milligram in one litre of water gives 1 ppm solution strength. The growth regulators may be applied in powder form or paste (lanolin paste use 200 g lanolin for 1 g chemical) or spray solution. Good water should be used for dissolving chemicals (ethylene and inhibitors). If the growth regulators is insoluble in water, like auxins (ethylene alcohol, methanol, potassium hydroxide, sodium hydroxide) and gibberellins (ethanol, methanol, 0.1N HCL) suitable solvents can be used to dissolve the chemicals.

It is a general rule that spraying of growth of growth regulators should be taken up in early morning or late evening hours for better utilization of the chemical. Highvolume hand operated sprayers are recommended for spraying. The prepared solutions should be kept in dark room and stored in refrigerator for 7-10 days.

a. **Talc method:** Some of the plant regulators are in powder or dust form. The cuttings may be moistened with water at their lower ends and then dipped in powder and planted afterwards, lower end beneath the soil eg. Seradix A andB can be used in the form of powder.
b. **Quick dip and prolonged soaking method:** Some of the chemicals are used in the form of solutions. They are dissolved in alcohol and then in

water. The lower end of the cuttings may be soaked in this solution. The concentrations ofthe solution vary from 10 to 2000 ppm. If the concentration of the solution is 10-50 ppm the cuttings may be soaked for 18 to 24 hours. (Prolonged soaking method) if it is 500 to 2000 ppm, the cuttings are soaked for a minute or less. This is called quick dip method. The concentration differs according to thetype of cuttings.

c. **Spray and paste method:** Growth regulators are used in aerosols form application in green houses for rooting in soft wood and herbaceous cutting. Mother plants spray such solutions (25 to 100 ppm) prior to taking cuttings from such plants for rooting enhancement. In paste method lanolin (wood fat, semi solid nature, yellowish in colour, by gentle heating can be made into liquid form mixed with powder from growth regulators and apply in air layering and stooling for root enhancement.

Crop	PGR	Dose	Time of application
Mango	NAA or 2, 4-D	20 or 10 ppm	Last week of April or when fruit attain marble stage
Citrus	2,4-D or Gibberellic acid	8-10 ppm or 50 ppm	Two spray at fruit development stage
Litchi	NAA or 2,4-D	10 ppm + 1% ZnSO4	At fruit development stage
Grape	Gibberellic acid	100 ppm	Before harvesting
Apple	NAA or 2, 4-D or 2,4,5-T	20 or 10 or 20 ppm	Immediately after petals drop
Bottlegourd	MH	50-150 ppm	Induction of more female flower

Crop	PGR	Dose	Time of application
Cucumber	GA_3	1500-200 ppm	Induction of more male flower
	Etheral	150 ppm	Induction of more female flower
Watermelon	TIBA	25-250 ppm	Induction of more female flower when spray at 2 leafstage
Gladiolus	Ethylene Chlorophydrin Thiourea GA_3 BA	4-5drops/litre 500ppm 50 ppm 25-50 ppm	Break dormancy of corm

To prepare solution conversion formula:

$$1ppm = mg / litre$$

$$1\% = \frac{ppm}{10000}$$

Exercise No. 18

OBJECTIVE –To Study about Herbaceous, Shrubbery Border and Plant Components of Garden

The planting of annuals in the border of the plot is called as herbaceous border. Dueto wide range of flowers available in winter it becomes a site of attraction in the garden. The concept of herbaceous border has been introduced in India gardens by britishers and now it has become a common feature.

1. The herbaceous border can be of two types:
2. **Single face herbaceous border:** It is made when border is situated on one side. Single face herbaceous border is made when border is situated on one side of plot and having background like wall or shrubbery border. Annuals are arranged in the border according to their height *i.e.* tall plants in back; mediumin center and dwarf in front line.

Double face border: It is more ideal when border is to be made in between a big plot. In such border, there is no background and tall annuals are planted in the center; medium and dwarf on both sides in descending order so that beautycan be watch from both sides.

Selection of site for herbaceous border: It is very important and desired picturesque effect may not be obtained if site is not selected properly. The site should be open sunny and facing towards south side so that plants get sufficient sunlight for maximum hours of the day. If this direction is not available, it should face east and last choice is for the west.

Back and Fore ground: To make effective herbaceous border there should be suitable background as well as fore ground. Back ground may be of dwarf flowering trees, shrubbery border or back ground made by training of sweet peas, brick wall in combination of both. Fore ground is generally of turf grass which may or may not have other garden ornaments like lily pool, sun dial, statue etc. Green turf in front of herbaceous border is a living carpet which serves a place for enjoying beauty of flowers while resting.

Size of the Border: The size of border depends upon the availability of space. The border can be of any length but width should be about 1-1.5 m or more depending upon the area of fore ground. Therefore, the width can be adjusted accordingly.

Colour Scheme of the border: The arrangement of these annuals according to colour scheme conveys the mood and feelings of the gardener. Dominance of a particular colour in border is suggestive quality of that colour. Hence, the arrangement of these annuals should be done very carefully. In nature green colour being the colour of foliage dominates in garden throughout the year. Warm colour like red dominates in winter whereas yellow colour in spring. There are three main colour scheme according to which these annuals can be arranged.

1. **Monochromatic colour scheme:** It restricts the use of single colour shades which maybe available in the same or different annuals. For example if blue colour is to be used annuals like blue corn flower, blue larkspur, ageratum, anchusa etccan be used. For yellow colour annuals like yellow dahlia, yellow antirrhinum,yellow calendula, yellow nasturtium, yellowmarigold, yellow annual chrysanthemum, and yellow coreopsis etc. can be used. The difficulty may be in getting the seeds of pure colour.
2. **Analogous colour scheme:** Annuals are arranged according to the nearest wave length of colour. This arrangement harmonizes with the surroundingbeds and hence it is called as harmonious colour scheme. The arrangement of colours will follow the order as white, creamish yellow, light yellow and finally deep yellow which will follow the same but in descending order.
3. **Contrast colour scheme:** The opposite colour of colour wheel are used and plants of contrast colours are planted accordingly e.g. red dahalia against the background green or next to blue corn flower creates a goo contrast. Foryellow colour violet, blue and red contrast. Yellow gamolepis red phlox or petunia and white alyssum can be planted for brilliant contrast. The width of the border is roughly divided into three parts in proportion of 5:3:2 or 6:3:1 fortall, medium and dwarf plants. Similarly, the plots are made curvaceous and thus plots become irregular in shape. In these plots planting is done irregularly at an approximate distance so that the border looks natural at the time of bloom.

PLANT COMPONENT GROWN IN GARDEN

1. **Annuals:** Commonly known as seasonal, constitute an amazing group of popular flowering ornamentals widely grown in all types of parks and gardens. Enchanting colours of the flowers in various shades, forms and fragrance have made them an obvious choice for all garden lovers.

a. ***Summer annuals (sown in Feb-March)***

- Kochia, zinnia (most beautiful summer annual), portulaca, gomphrena, sunflower

b. ***Rainy season (sown in June-July)***

- Balsam, cock's comb (celosia), gomphrena

c. ***Winter Annuals (Sown in September –October)***

- Antirrhinum, China aster, Cornflowers, Larkspur, Sweet sultan, Phlox, Verbena, Candy tuft, Petunia, etc.
- Winter annuals produce white flower: Candytuff, stock, aster, phox etc.
- Winter annuals produce blue flower: Anchusa, cornflower, Ageratum etc.
- Winter annuals produce red flower: Phlox, poppy, godetia, lupin
- Winter annuals produce yellow flower: Calendula, Marigold, sunflower,coreopsis
- Fragment flower: stock, sweet pea, sweet sultan
- Shady corner: Salvia, cineraria
- Planted in hanging pots: Dasiy, nasturtium, verbena, phlox
- Dry flower: Statice, Helichrysum, Acroclium, Nigella, lady's lace
- Edges: Phox, verbena, pansy, candytuff etc.
- Rock garden: Ice plant, verbena, phox, verbena
- Cut flower: Carnation, sweetpea, sweet William and Larkspur
- Loose flower: Marigold & Annual chrysanthemum
- Pot flower: Antirrhinum, Aster, petunia and carnation

SOME UNIQUE ANNUALS KNOWN FOR THEIR BEAUTY ARE AS FOLLOWS

- **Kochia:** Summer leafy annual which has columnar shape hence named as summer Cyprus and in the end of season slowly it change to brown colour looks like burning hence named as 'burning bush'.
- **Salvia and Cineraria:** Shady annuals
- **Sweet pea:** It is a climber and used for screening.
- **Holyhock:** Very tall so widely used for screening purpose.
- **Calendula:** yellow in colour, also known as pot marigold
- **Antirrhinum:** Flowers exhibit like barking dog when adjacent sides are pressed hence commonly known as 'snap dragon, dog flower, bunny flower, rabbit flower'.
- **Clianthis:** Peculiar flower shape having parrot beak shape.

2. **Trees:** They are fascinating because of their graceful appearance and the abundance of bloom. They are grown for their economic importance value or both. The cultivation of trees for their aesthetic value is known as **arboriculture**. Trees should be planted carefully and thoughtfully for the benefit of height, shade, colour and vertical emphasis.

a. **Trees for screening purpose:** Silver oak, Popular, Ashoka, Eucalyptus
b. **Trees for checking pollution:** Ficus spp.
c. **Tree for alkaline and saline soil:** Amaltas, *Casurina equistifolia.*

Description of Some Beautiful Trees

Common Name Scientific Name	Remarks
Silver oak (*Grevillea robusta*)	Common screening and shading tree used in coffee plantation
Golden rain tree (*Koelreutaria panniculata*)	It is very beautiful tree having leaves like neem which comes in yellow colour flowering in month of September-October and then flowers transform into reddish-orange colour trifid fruits in October-November (flowers welcome the winters in north India)
Semal or Red cotton tree (*Bombax malabricum*)	Towering trunk having buttress roots produce attractive reddish-orange color flowers in month of March-April (spring season)
Mexican cotton tree	Beautiful pink colour flowers come in October-November (flowers welcome the winters in North
(*Chorisia speciosa*)	India). Leaves shed during flowering time and tree looks very romantic and magnificent.
Flame of forest or Dak (*Butea monosperma*)	Crescent shape orange colour flowers bear on tree in March-April. Flower looks like burning candle as very showy
Bottle brush (*Callistemon lanceolatus*)	Bottle brush shape red flowers come in month of March-April.
Amaltas or Golden laburnum (*Cassia fistula*)	This is very unique tree which is adapted to saline soils and holding great aesthetic beauty. In harsh months of May-June it bears hanging yellow flowers which are eye catching hence named golden laburnum"
Red gulmohar or Peacock flower (*Delonix regia*)	Leaves are very beautiful and red orange colour flowers in May-June cover whole tree
Nili Gulmohar (*Jacaranda mimosifolia*)	Impressive tree with purple-blue colour flowers bear in May
Kadamba (*Neolamarckia cadamba*)	Orange coloured, round ball shaped flowers which are fragrant.
Pink Gulmohar (*Cassia javanica*)	Dashing pink crimson colour flower comes in spring
Pride of India (*Lagerstroemia speciosa*)	Outstanding summer bloomer bear flowers in mauve and pink colours. L.indica commonly known as Swani (rainy season) is a shrub and small form of this tree
Temple tree (*Plumeria alba, P. rubra* and *P. pudica*)	*P. pudica*-Evergreen tree bears white colour and non-fragrant flowers *P. alba*- Deciduous trees bear intensely fragrant white flowers with yellow centre in June-November. *P. rubra*-Winter deciduous plants bear red colour flowers

Common Name Scientific Name	Remarks
Tree of heaven (*Ailanthus excels*)	Popular avenue tree
Fountain tree or African tulip tree (*Spathodea campanulata*)	One of the world's spectular flowering trees with glorious orange scarlet flowers. It may grow to 80 ft on an ideal site, but most specimens are much smaller. The tree flowers through the growing season, but peak bloom is usually in the spring
Kanak champa (*Pterospermum acerifolium*)	Coarse textured leaves and tree bears yellow colour flowers in April-June.
Weeping willow (*Salix babylonica*)	Charming ornamental tree, widely cultivated for its weeping branches which give a romantic effect, especially when the tree is on a water bank. The flowers are arranged in catkins produced early in the spring. Weeping willow has male and female flowers on separate trees. Weeping willow is extensive in cultivation all over the world.
Akash neem/Cork tree/Tree jasmine (*Millingtonia hortensis*)	A profusion of silvery-white, delightfully fragrant, flowers starts from April until the rains and again in November and December.
Ficus species e.g. *infectoria* (pilkhan), *retusa* (chilkhan), *religiosa* (papal), *F. benghalensis* elastic (rubber)	Ficus species belongs to moraceae family which is shady trees and mostly used for bonsai purpose except rubber as it has large leaves.
Traveler's tree or Dancing peacock (*Ravenala madagascariensis*)	Banana like leaves bear at the top of trunk in fashion which seems peacock is dancing hence common named dancing peacock. It store water in between the leaves which travelers used in past while commuting as commonly named traveler's tree and originated from Madagascar that is why botanically so called.
Moulsari (*Mimusops elengi*)	Circular, most dense canopy very suitable for shade purpose.
Maple *Acer* spp.	One of the most elegant, classical and romantic tree species, equally loved by poets, lovers and naturalists. This is insignia to Canadians. Maple leaf on the army dress and national flag of Canada shows importance. This is most widely used bonsai lovers. Some species changed their colours to various to various shades of yellow, pink, red and others are deciduous nature which leads to colouring carpet of leaves it magical tree and awe-inspiring.

3. **Climbers-** These are week woody plants which climb with support of special modifiedorgans they possessed.

- **Tendril:** *Antigonon leptopus* (coral vine), *Bignonia gracillis*, *Pyrostegia venusta* (golden shower)

- **Thorns:** Bougainvillea, rose climbers
- **Sticky roots:** Ficus spp.

CLASSIFICATION OF CLIMBER BASED ON UTILITY

a. **Annual Climber:** *Clitoria ternatea*, sweet pea, morning glory (*Ipomoea rubrocaerulea*)

b. **Screening:** *Pyrostegia venusta* (golden shower), *Vernonia elaeagnifelia* (perda bel or curtain vine)

c. **Indoor climber:** Pothos (Money plant), monster, philodendron, asparagus

d. **Sunny climbers:** *Pyrostegia venusta* (golden shower), *Antigonon leptopus* (coral vine), *Adenocalymma alliaceum* (garlic vine), *Quisqualis indica* (Rangoon creeper or jhumka bel)

e. **Partial Shady climbers:** *Clerodendron splendens* (*C. inermi* is a shrub), *Petrea volubilis* (purple wreath), *Lonicera japonica* (Japanese honey suckle)

f. **Heavy climber:** *Hiptage benghalensis* (madhvi lata)

Description of Important Climbers

Garlic Vine *(Adenocalymma alliaceum)*	Sun loving climber bears elegant lavender or purple coloured flowers in autumn (October-November), leaves smells like garlic on crushing hence named.
Coral vine *(Antigonon leptopus)*	Sun loving climber bear beautiful pink coloured flowers in September-November, leaves are heart shape, climb through tendrils.
Clerodendron splendons	Shade flowers loving climber produce heavy red colour
Ficus ripens (Indian ivy)	Stick to walls with sticky rootlets, commonly used in wall gardening
Loncera japonica (Japanese honey suckle)	White colour flowers
Pyrostegia venusta (Golden shower)	Impressive climber bears orange coloured trumpet shaped purposes and in pergola making
Petrea volbilis (Purplewreath)	Dazzling purple colour flowers in form of wrath, leaves have coarse texture like sand paper
Vernonia elegnaefolia (Perda bel or curtain vine)	Elegant hanging climber like curtains apparently used in screening
Wisteria sinensis (Grape flower vine)	One of the most beautiful climber bear handsome mauve coloured flowers in bunch in March, common in temperate and cool regions.
Lathyrus ordoratus(Sweet pea)	Annual climber apparently used for screening and vertical gardening
Quisqualis indica (Rangoon creeper or Jhumka bel)	Pink-reddish colour throughout the year flowers come in cluster

Hiptage bengalensis (Madhavi lata)	Heavy climber bears white flowers in February-March.
Passiflora laurifolia (Watch flower)	Watch shape orange colour flowers comes in summer.
Jasmine spp.	Fragrant white flowers widely used for perfume extraction (*J. auriculatum* and *J. grandiflorum* are commonly used for perfume extraction.
Ficus spp. (*repens* and *pumila)* (Crepping fig/climbing fig)	Popular climber for wall gardening or vertical gardening. Propagated by simple layering and stem cutting required partial sun-light

4. **Shrubs-** Wood plants are generally planted in triangular fashion in double row. They are main filler plants of the gardens and have wide range of variation with regard to growth, leaf texture and flower (shape, size and colour). They bloom all round theyear and help gardens presentable.

Shrub	Remarks
Mussaenda	The plant's color comes from large bracts often yellow, white, red or orange with flower at the center of each bract. Bracts may be seen in several colours including rose, white, red, pale pink and some mixtures
Bougainvillea	Belong to family Nyctaginaceae. It is a sun loving semi-climber and can be grown as a hedge, a shrub, a climber over a sunny wall and also in pots
Poinsettia or Christmas plant (*Euphorbia pulcherima*)	Dashing coloured bracts (modified leaves) are reasonof this beautiful plant
Weeping merry (*Russelia juncea*)	Dwarf shrub bear very beautiful red flowers
Tree of sadness or Harsingaror Night jasmine (*Nyctanthus arbotristis*)	Bear fragrant white-orange tinge flowers in night which shed in the morning hence common named,tree of sadness
Peacock flower or Garden Gulmohar (*Caesalpinia pulcherrima*)	Bear orange colour flower which look like gulmohar.
Scarlet bush (*Hamelia patens*)	Leaves bear in circular fashion with red flowers
Chinese shoe flower or Gurhal (*Hibiscus rosa sinensis*)	White-creamy catkin inflorescence
Parrot's beak (*Gamelina hystrix*)	Elegant shrub bears fascinating yellow coloured flowers resembling parrot's beak merge from a pendant structure of overlapping bracts.

5. **Hedge** is an important garden features both for beautification and functional features. When selected plant species is planted together at close distance, in single or multiple rows, later pruned to retain a definite shape and height is known hedge. These are actually live fences used for demarcating road, paths and sometimes for providing physical barrier, screening and several other purposes.

Selection of Plants

Foliage plants: These plants have either green or variegated leaves and generally preferred for the development of hedge

- *Acalypha wilkesiana'Tricolor', A. wilkesiana'Marginata', A. 'Batique'*-green and variegated leaves
- *Bryenia nivosa, Bivosa'Roseo-picta'*- ornamenral variegated foliage, suitablefor tall hedge.
- *Casuariana equisitaefolia*– ideal for tall and wide hedge.
- *Clerodendrum inerme*– Leaves, compact growth, very popular, suitable forlow height hedge.
- *Duranta plumieri'Variegata'*– Elegant, variegated leaves, ideal for low heighthedge.
- *Polyscias spp/cvs*– Having both green and variegated leaves suitable for semishady locations.
- *Lawsonia inermis*– Leaves green, small, uncommon but suitable fordeveloping wide hedge.
- *Pedilanthus tithymaloides*– Leaves variegated suitable for shady location.

Flowering Type– Several flowering shrubs are used as hedge.

- *Ixora bandhuca, I. coccinea, I. stricta*– Hardy and excellent, suitable for short length.
- *Euphorbia milli*– Thorny and bears small flowers also, suitable for protectivehedge.
- *Murrarya exotica*- Produce fragrant flowers besides shining dark green leaves,compact, and ideal for small temple gardens.
- *Barlaria cristata*- Thorny and produce flowers in white, mauve, suitable forsunny locations and small gardens.

Planting: For developing wide hedge having 50-60 cm width, the trench should be of 60cm x 60 cm in depth and width and the length runs as per requirement.

Pruning: It is essential operation in order to keep the hedge in shape. Frequency of pruning is generally dependent on type of the plant species and season. During summer and rainy season hedges are required to be pruned frequently in comparisonto winter. However, the ultimate aim of pruning is to keep the hedge in proper shape.

Shapes: Flat top most popular and common. Other fascinating shapes like wavy, zigzag etc are given to make hedge more attractive and purposeful. There is nodefinite guideline about shape but it should be elegant and well integrated to the overall design of the garden.

6. **Edge** is mainly used to demarcating beds, lines and subdivision of the area so that it looks more elegant and used exclusively for aesthetic purpose. Edge is also developed in the same fashion like hedge but it is of

much low height and width. Height of the edge should not be more than 15 to 25 cm from the ground level. For developing edge, a narrow trench 30 x 30 cm is dug at desired place, thoroughly prepared, manures like hedge and rooted cutting/plants are planted at the distance of 15-25 cm depending upon the plant species in single row.

Following plants are suitable for this purpose: *Aerua sanguinolenta, Alternanthera* spp. *Chlorophytum* spp., *Duranta repens, Duranta* 'Golden',*Eupatorium foeniculifolium, Iresine herbstii, Ohiopogon* sp, *Setcreasea purpurea* etc.

Topiary: The word 'Topiary' has been derived from latin word 'topiarius', the ornamental landscape gardener and the creator of 'topia' or places, referring a particular indoor gardens. Topiary, in Horticulture, is a practice of training live perennial by pruning and clipping the twigs and foliage in order to give definite shape over period of time. The design and shapes may be geometric or any fancy thing outof one's imagination. Usually various shapes may be geometric or any fancy thing outof one's imagination. Usually various shape of animals, birds or any geometric shapes/forms are made as topiary.

Selection of Plant species: Plants selected for topiaries have following characteristics:

- Evergreen
- Woody perennials
- Leaves-small or needle like
- Dense foliage
- Compact growth
- Response to pruning, clipping and pinching

In sub-tropical/tropical climate conditions, following plant species are suitable for this purpose- *Casuarina equisetifolia, Duranta plumier, Duranta* "Golden', *Hibiscus liliflorus, H. liliflorus 'Variegatus', Juniperus chinensis, Murraya excotica, Polyalthia longifolia* 'Pendula', *Putranjiva roxburghii, Thuja orientalis.*

7. **Bonsai** is combination of two Japanese words Bon (meaning shallow pan) andSai (Meaning plant), which deciphered as tray planting. It is a fascinating Japanese artborn in China.

STYLES OF BONSAI

- **Upright or Chokkan style:** In this style a single specimen is grown with a plant straight and upright stem.
- **Winding or Kyokkuk Style:** Single trunk is twisted once or several times, showing as if it has struggled for its existence in the projection to wild.
- **Oblique or Shakan style:** This referred to as windswept style. Plant is grown in a slanting position exhibiting plant has been swept by wind.
- **Cascade or Kengai style:** Trunk hangs over the edge of container exhibiting cascading effect.

- **Ikadi-Buki style:** A tree is trained horizontally and several branches are allowed to grow vertically at intervals resembling individual trees. It gives resemblance of forest.
- **Clasped to stone type:** a single tree is grown with its roots trained over stone or some blocks.
- Pruning, pinching, wiring and repotting are important operations to be performed in Bonsai development.

9. **Carpet bedding:** Bedding is an important feature in landscape gardening and creates a great impact by virtue of their colour combination and attractive design. Carpet bedding refers plantation of fast growing flowering and foliage plants inparticular design in different season. The main purpose is to create novelty with the help of flower and plants and to make the gardening interesting.

Types: Carpet bedding is created in different season like winter, spring and summer accommodating plants of that particular season. Seasonal bedding provides scope of variation in design and flower colour. Some carpet bedding is made as perennial feature particularly when monograms and institute/ company name is inscripted with plants.

Design: It should be simple but attractive following colour scheme depending upon the season and purpose.

Colour combination: The attractiveness of the design is main criterion and that created by proper colour scheme. Selection of plant species is to make dependingupon the colour scheme.

Selection of Plants: The plants usually selected for carpet bedding are herbaceous, annuals/biennials or perennials. Both foliage and flowering types are selected as per suitability of the design.

Foliage: *Casuarina equisetifolia, Duranta plumier, Duranta* "Golden', *Hibiscus liliflorus, H. liliflorus 'Variegatus', Juniperus chinensis, Murraya excotica, Polyalthia longifolia* 'Pendula', *Putranjiva roxburghii, Thuja orientalis.*

Flowering: Brachycome, Candytuff, French marigold (dwarf), Ice plant (*Hymenantherum, Mesembryanthemum),* Nasturtium, Pansy, Phlox, Portulaca, Sweet Alyssum,Verbena.

Exercise No. 19

OBJECTIVE – Lawn Management

Lawn is a ground cover of perennial grass, which persist in close mowing and requires proper management practices.

PURPOSE OF LAWN

- It is an important element in the garden
- It leads to unity in garden design.
- It is a natural green carpet and is the carpeted floor of outdoor room.
- It is the heart of the garden and the centre of social life
- It is the centre piece around which all the garden elements are placed in sub- ordinate order like royal court, where king occupies the central position and is surrounded by the his courtiers.
- It gives restful appearance to the eyes through its green outlook all the time.
- A lush green lawn is refreshing especially during summer.
- Lawn is the best foreground to enjoy the charm and beauty of the ornamentalplants and features.
- Prevent soil erosion

CHARACTERISTIC OF A LAWN GRASS

1. Look fresh and green throughout the year
2. Not patchy
3. Cold or drought resistant
1. Free from attack of diseases and insects
2. Quick growing

3. Soft to touch
4. Not giving fowl or bad smell

Ideal Site for Planting Lawn

- South, south east or south-west open and sunny place for most part of the daywith adequate water availability.

How Leveling and Grading of Ground for Lawn is Done

- Level the soil for uniformity of growth throughout the entire area
- Do perfect grading for proper drainage i.e one foot in every 50 feet.
- Leveling is checked visually, flooding the area, stretching the rope and avenuelevel.

Characteristic of Ideal Soil and Method of Preparing Soil for Growing Lawn:

- Sandy-loam, well fertile, drained with good water holding capacity having pHof 5-6 and sufficient humus or organic matter.
- Dig soil up to 45 cm depth and expose to sun in May-June.
- Turn soil 2-3 times, remove stones, rocks and break big clods
- Spread 10-15 cm thick layer of well rotten weed free farm yard manure andthoroughly mix in soil.
- Irrigate the field thoroughly and allow weeds to geminate.
- Remove all the weeds along with roots manually or spray non-selective typeof herbicide like paraquat or gramaxone @ 1.0-1.5 litre per hectare in about 800-1000 litre water.

METHODS OF LAWN RAISING

1. Seeding

- This method is common to grow cool season lawn grasses.
- About 25 kg seed is mixed in 200-250 kg sand or saw dust and is broadcasted evenly in the prepared field.
- Do light rolling
- Sprinkle water regularly until seedling emerges.
- Less labour is required, but lawn is not even.

2. Dibbling

a. A small bunch of grass along with roots and little stem is taken
b. Planting is done at a spacing of 10 cm apart both row to row and plant to plant.
c. Do regular watering until establishment.
d. Dibbling is done in June to September.
e. Lawn develop by this method is quick, uniform and with more labour and cost.

3. Plastering

a. Grass roots with little stem of 4-5 cm long pieces are mixed with compost and Farm yard manure.
b. Spread this over prepared field during rainy season.
c. Avoid heavy balling.
d. Do liberal watering with sprayer.
e. Do mowing with sprayer.
f. Do mowing after 70-80 days

4. Sprinkling

a. Grass roots along with little stem are chopped into small pieces.
b. Spreading this over prepared field during rainy season.
c. Do small raking to mix grass in soil.
d. Do light rolling.
e. Do liberal watering with sprayer.
f. Do mowing after 70-80 days.

5. Turfing

a. Small pieces of well prepared lawn or turf are cut into square orrectangular shape preferably planted on polythene sheet.
b. Fix these in a thoroughly prepared field.
c. Do heavy rolling.
d. Lawn prepared is clean and weed free.
e. Quickest method of lawn raising

TYPES OF GRASSES

A. Bermuda Grass (Cynodon Dactylon)

- It is commonly called as doob grass.
- Very commonly used for planting lawn due to its fast growth, hardiness and lesswater requirement.
- Response to frequent mowing.
- Makes an excellent turf.
- Highly suitable for large areas and play ground.

B. Korean Grass (Zoysia Japonica)

- Native of Japan, Korea and Philippine islands.
- Very popular during short span of time due to its velvety growth and morecold tolerance.
- Tough grass difficult to mow.
- Lawn should be 1.5 -25 cm height.

C. Bahia Grass (Pospalum Notatum)

- It is not an aggressive spreader but is a drought resistant turf and requires noirrigation.
- It does not require excessive fertilization.
- When fertilizer is applied, it should contain iron, especially if the soil pH is 7or more.
- It is best used in sunny areas in warm humid regions. Its roots can extend up to 8 inch deep.
- It can tolerate moderate shade and thrives well along highways.

D. Kentucky Bluegrass (Poa Pratensis)

- Common cool season grass which gives high quality lawn and is available inblends.
- Bluegrass develops shallow roots.
- It growth slows during the warm summer months.
- It is susceptible to disease and weed invasion.
- It requires low mainatence and poor to shade tolerance.

E. Carpet Grass (Axonopus Fissifolius)

- It is also known as flat grass.
- It is a perennial, coarse leveled, creeping grass, flat stolons and a tall seed stalk with 2 branches at the apex.
- It will grow well in either sun or shade.

F. St. Augustine Grass (Stenotaphrum Secundatum)

- Also known as Charleston grass.
- It is native to the Caribbean, Africa and Mediterranean regions and best adapted to subtropical climates.
- It is susceptible to fungal diseases.
- Texture is coarse with poor cold and traffic tolerance with moderate shade tolerance.
- It has medium to fast establishment rate.

Mowing and its Effect

It is the cutting of lawn grass for maintaining its attractiveness for minimum utility

1. Mowing stimulates bud development
2. Shoots become thicker and roots shorter.
3. Generally lawn height is maintained at 5-7 cm, as very close mowing results in week growth and mowing at more height may not serve the purpose.
4. In one mowing not more than $1/3^{rd}$ leaves are removed.

5. Remove stones or pebbles before mowing, so as to avoid any damage to shersof lawn mower.
6. In most of the grasses mowing is done at fortnightly interval.

Rolling

1. Uniformity of growth is achieved by rolling, as it touches the nodes at ground,thereby keeping the area well leveled.
2. Heavy rolling is done at the beginning of rainy season followed by lightrolling in subsequent months depending upon grass species.

Fertilizer Application

1. Sunhemp is very good green manure before lawn planting.
2. In a 30 m^2 lawn apply 3-5 q well rotten FYM, 10-20 kg lime and 10-20 kgSSP.
3. Broadcast a mixture of 50-60 g/m^2 (2CAN: 1SSP: 1 K_2SO_4) twice inFeb-ruary-March and August-September.
4. Spray of urea 0.3 per cent is also beneficial.

Irrigation

1. Frequency and amount of irrigation depends upon soil, grass, weather andclimate.
2. It should be done before wilting or internal water stress.
3. Increased watering result in deeper root development, thereby increasing waterrequirement.
4. Watering up to 15 cm depth at 8-10 days interval is ideal as frequent lightwatering is harmful.

Weeding

1. Regular mowing checks weed growth by removing upper plant portion andstarving roots.
2. For small area hand weeding is done.
3. For controlling broad leaved weeds spray 2,4-D (0.05%)
4. For narrow leaved weeds spray Atrazine @1.5 a.i. per hectare in 1000 litre water.
5. In lawn of Korean grass spray benefin or sylvex (0.1%)

Precautions

1. Avoid water standing in rainy season.
2. Remove dead or dry leaves falling during autumn season.
3. Do raking in lawn twice once before rains and second after rains.
4. Do thinning as and when required

Exercise No. 20

OBJECTIVE – Irrigation Management

Irrigation is the artificial application of water to land for growing crops or trees. According to Israelsen and Hansen (1962), the artificial application of water for the purpose of supplying moisture essential to plant growth is called Irrigation

METHODS OF IRRIGATION

A. Surface Irrigation

Apply water to the soil without aerial application is known as surface irrigation. Different systems of surface irrigation are as follows:

i. **Flooding:** This is followed in wet lands mostly for banana. This is a wasteful method which will lead to stagnation of water and help weed growth.

ii. **Check:** Check bunds for large areas enclosing a number of trees are provided with channels between two rows. This is more economical than flood system.

iii. **Basins:** This is widely practiced. The basins should be square or circular and should be sloping from the trunk to the periphery. This method is useful in young orchards, light sandy and alkaline soil. The size of the basin should be widened as the roots spread.

iv. **Ring:** In this system, small ring bund will be provided around the trees or one single irrigation channel connecting all trees will be formed and around the channel is widened to form basin.

v. **Bed:** This is adopted in heavy soils for fruit crops like banana, wherein 3-4 plants are enclosed in a bed and is irrigated by opening in one side of the bed.

vi. **Furrow:** This is most widely followed for vegetable crops like tomato, onion, brinjal etc.

All the above different systems of surface irrigation do not ensure uniform distribution of water. It may be more near channels and less away from channels.

B. Sub Surface Irrigation

This method supplies water from below soil through underground pipes or by ditches on one side. This is useful for green houses. Pipes are laid 45-60 cm deep an 6 m apart. Pipes will have holes at regular intervals. This method is costly and deepcultivation is not possible. But evaporation of moisture is prevented to a great extent.

C. Special Irrigation Methods

1. **Overhead Irrigation:** Overall irrigation is by the use of sprinklers, mostwidely used over head system. In this system, the initial cost of installation is rather high but there are several advantages. There is saving in labour cost and water. More uniform wetting of soil is possible and erosion will be eliminated.This method is best for steep and terraced lands. This is more widely adopted in plantation crops. There are also some disadvantages. Due to the influence ofwind, there will be non-uniformity in coverage. In hot sun, droplets on leaves and fruits may cause sunburn. Certain diseases may spread easily.

2. **Drip irrigation:** Drip irrigation is known by various names like 'trickle irrigation' or 'high frequency irrigation' or daily flow irrigation. This is a method of watering plants at a rate of equivalent to its consumptive use, sothat plants would not expensive any stress during the growing phase. In thisthe water is conveyed from a source under low pressure to the root zone of the crop only. The twin objectives of this method of irrigation are (1) provision of optimum quantity of water to the crop for optimum production and (2) saving the valuable water from wastage thereby increasing the water use efficiency and the command area.

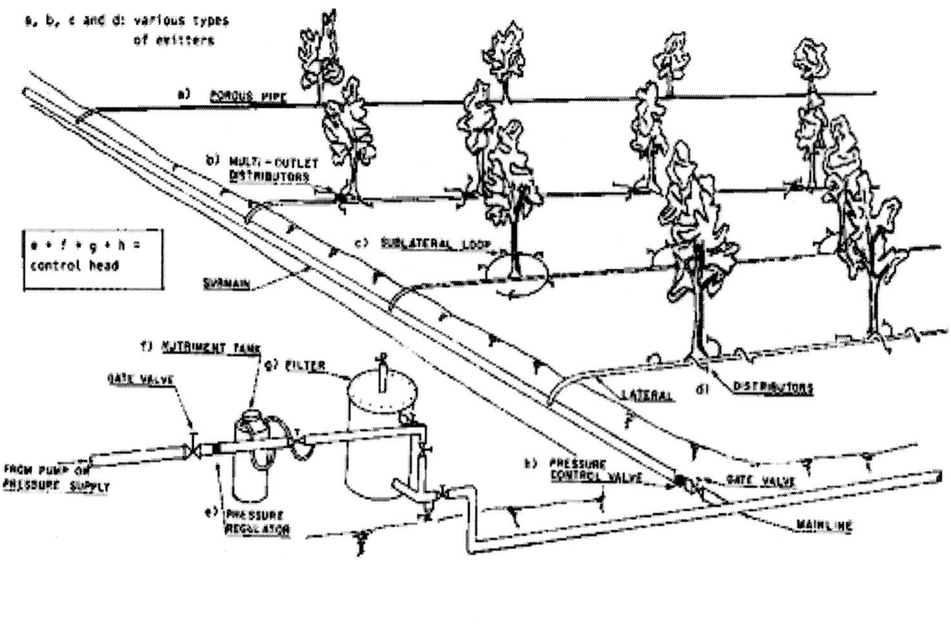

Water requirement: Pan Evaporation method is used for estimating crop waterrequirement.

$$V = \frac{CA\left(m^2\right) x PE x P_c x K_c x PWA}{E_u}$$

Where, V =Volume of water required (i/day/plant)

CA = Crop area (m^2) or spacing

PE =Maximum pan evaporation (mm/day)

Pc =Pan coefficient (0.7 to 0.9)

Kc =Crop coefficient (value depends on density 0.5 to 0.8)

PWA= Percentage wetted area (depending on age of fruit tree)

Eu =Emission uniformity (0.9)

Fertigation or herbignation: Fertigation is the application of fertilizers, soil amendments or other water soluble products through an irrigation system. Final fertilizer mixing water should have pH of 5.5 to 6.5, EC 0.1, low carbonate (2.25), boron (1ppm), sodium (SAR 15), chloride (5ppm) and nitrate (15 ppm) and free from heavy metals.

Success of fertigation depends upon (a) application of nutrients in the active feeder root zone of plants in accordance to their relative requirement for different nutrients atdifferent phonological growth stages (b) selecting the sources of fertilizers with reference to the soil physcio-chemical characters, water quality, the need for combining the fertilizers and their compatibility (c) calculating the quality of fertilizers per day for each growth stage and (d) allowing the dissolved fertilizers through drip system every alternate days or once in three days depending upon the filtration rate of the soil. Fertigation studies conducted with fruit crops like banana revealed that fertigation with 75-100 per cent of recommended dose of fertilizer at weekly interval results in increased bunch weight, quality of fruits etc. Similarly in guava 75% of recommended dose of water soluble fertilizer (WSF) with 6-10 split from fruit set and apply irrigation at alternate day gave better quality fruits.

Exercise No. 21

OBJECTIVE – Study of Different Types of Greenhouses Based on Shape

Green House: A greenhouse is a framed or an inflated structure covered with a transparent or translucent material in which crops could be grown under the conditions of at least partially controlled environment and which is large enough to permit persons to work within it to carry out cultural operations.

GREENHOUSE TYPE BASED ON SHAPE

Greenhouses can be classified based on their shape or style. For the purpose of classification, the uniqueness of the cross section of the greenhouses can be considered as a factor. As the longitudinal section tend to be approximately the same for all types, the longitudinal section of the greenhouse cannot be used for classification. The cross sections depict the width and height of the structure and the length is perpendicular to the plane of cross section. Also, the cross section provides information on the overall shape of the structural members, such as truss or hoop, which will be repeated on every day. The commonly followed types of greenhouse based on shape are lean-to, even span, uneven span, ridge and furrow, saw tooth and quonset.

1. Lean-to Type Greenhouse

A lean-to design is used when a greenhouse is placed against the side of an existing building. It is built against a building, using the existing structure for one or more of its sides. It is usually attached to a house, but may be attached to other buildings. The roof of the building is extended with appropriate greenhouse covering material and the area is properly enclosed. It is typically facing south side. The lean-to type greenhouse is limited to single or double-row plant benches with a total width of 7 to 12 feet. It can be as long as the building it is attached to. It should face the best direction for adequate sun exposure. The advantage of the lean-to type greenhouse is that, it usually is close to available electricity, water, and heat. It is a least expensive structure. This design makes the best use of sunlight and minimizes the requirement of roof supports. It has the following disadvantages: limited space, limited light, limited ventilation and temperature control. The height of the supporting wall limits the potential size of the design. Temperature control is more difficult because the wall that the greenhouse is built on, may collect the sun's heat while the translucent cover of the greenhouse may lose heat rapidly. It is a half greenhouse, split along the peak of the roof.

2. Even Span Type Greenhouse

The even-span is the standard type and full-size structure, the two roof slopes are of equal pitch and width. This design is used for the greenhouse of small size, and it is constructed on level ground. It is attached to a house at one gable end. It can accommodate 2 or 3 rows of plant benches. The cost of an even-span greenhouse is more than the cost of a lean-to type, but it has greater flexibility in design and provides for more plants. Because of its size and greater amount of exposed glass area, the even-span will cost more to heat. The design has a better shape than a lean-to type for air circulation to maintain uniform temperatures during the winter heating season. A separate heating system is necessary unless the structure is very close to a heated building. It will house 2 side benches, 2 walks, and a wide center bench. Several single and multiple span types are available for use in various regions of India. For single span type the span in general, varies from 5 to 9 m, whereas the length is around 24 m. The height varies from 2.5 to 4.3 m.

1. Uneven Span Type Greenhouse

This type of greenhouse is constructed on hilly terrain. The roofs are of unequal width; make the structure adaptable to the side slopes of hill. This type of greenhouses is seldom used now-a-days as it is not adaptable for automation.

2. Ridge and Furrow Type Greenhouse

Designs of this type use two or more A-frame greenhouses connected to one another along the length of the eave. The eave serves as furrow or gutter to carry rain and melted snow away. The side wall is eliminated between the greenhouses, which results in a structure with a single large interior, Consolidation of interior space reduces labour, lowers the cost of automation, improves personal management and reduces fuel consumption as there is less exposed wall area through which heat escapes. The snow loads must be taken into the frame specifications of these greenhouses since the snow cannot slide off the roofs as in case of individual free standing greenhouses, but melts away. In spite of snow loads, ridge and furrow greenhouses are effectively used in northern countries of Europe and in Canada and are well suited to the Indian conditions.

3. Saw Tooth Type Greenhouse

These are also similar to ridge and furrow type greenhouses except that, there is provision for natural ventilation in this type. Specific natural ventilation flow path develops in a saw- tooth type greenhouse.

4. Quonset Greenhouse

This is a greenhouse, where the pipe arches or trusses are supported by pipe purling running along the length of the greenhouse. In general, the covering material used for this type of greenhouses is polyethylene. Such greenhouses are typically less expensive than the gutter connected greenhouses and are useful when a small isolated cultural area is required. These houses are connected either in free, standing style or arranged in an interlocking ridge and furrow. In the interlocking type, truss members overlap sufficiently to allow a bed of plants to grow between the overlapping portions of adjacent houses. A single large cultural space thus exists for a set of houses in this type, an arrangement that is better adapted to the automation and movement of labour.

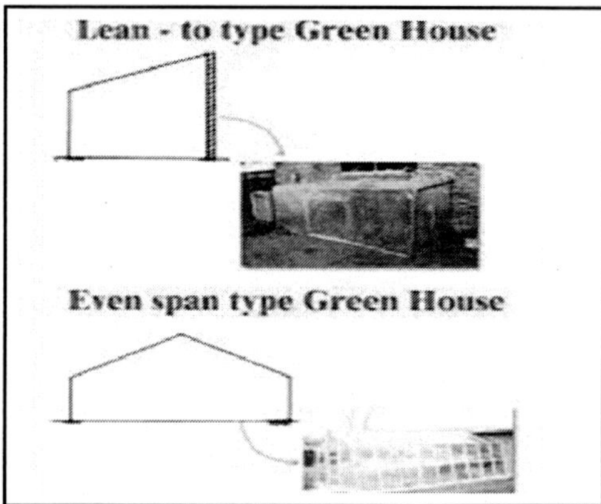

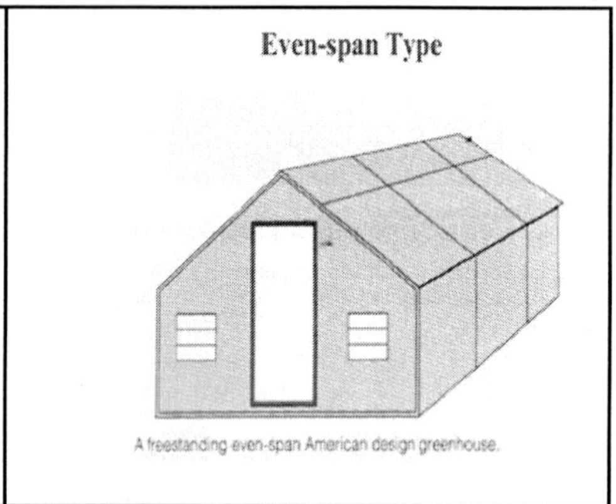

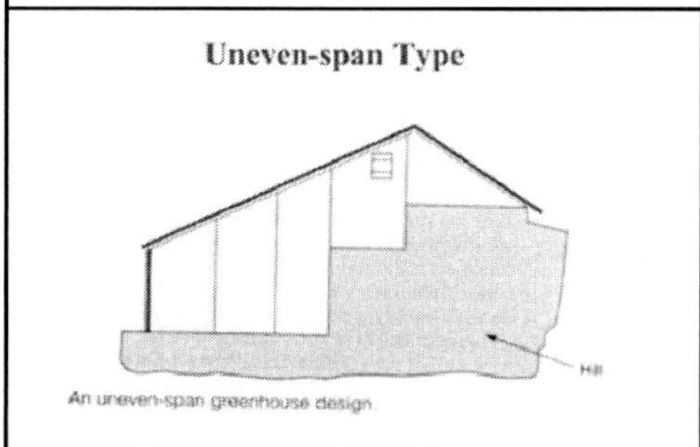

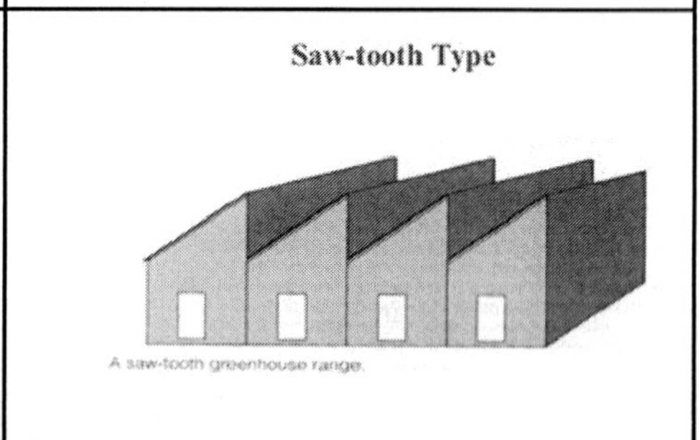

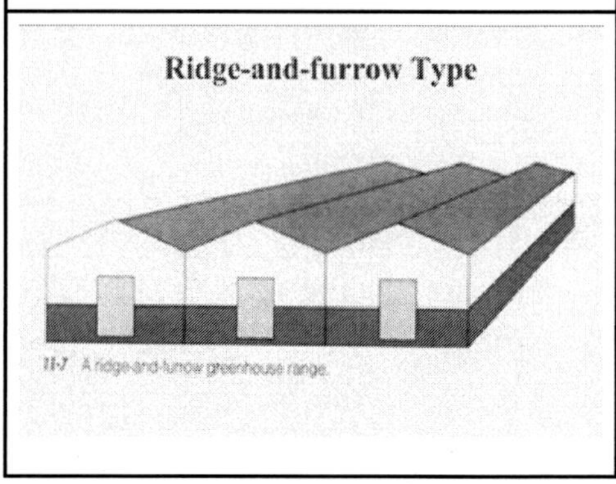

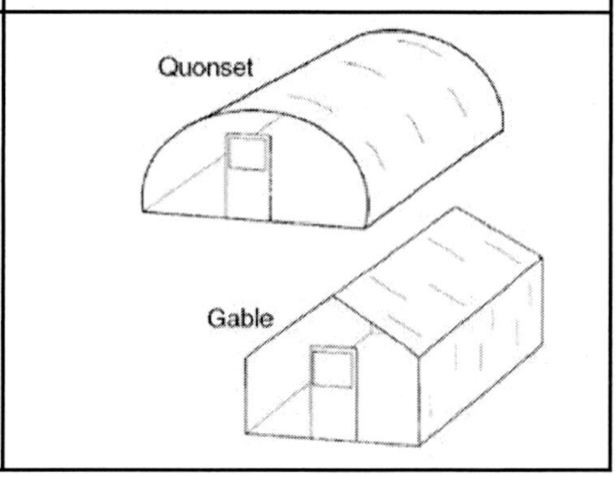

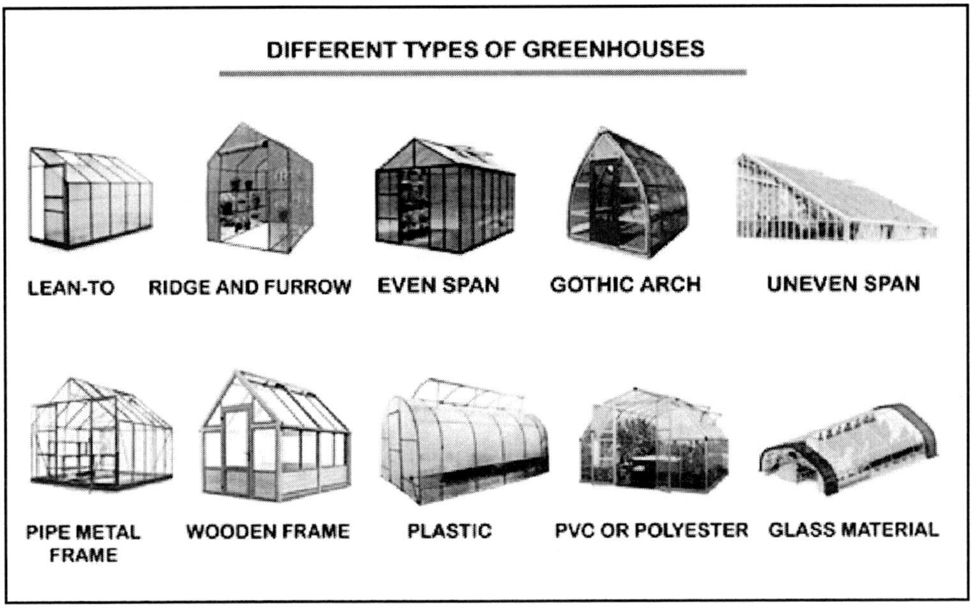
DIFFERENT TYPES OF GREENHOUSES
LEAN-TO
RIDGE AND FURROW
EVEN SPAN
GOTHIC ARCH
UNEVEN SPAN
PIPE METAL FRAME
WOODEN FRAME
PLASTIC
PVC OR POLYESTER
GLASS MATERIAL

Exercise No. 22

OBJECTIVE – Study on Hydroponics

HYDROPONICS

Hydroponics is the process used to grow plants without soil and literally means 'working water'. The grower is taking 'control' of the plant's root environment, and losing the benefit of 'mother nature's' finely-tuned mechanisms which normally control that part of the plant's environment. Hydroponics is not an easier way to grow plants! It is a more controlled way of growing plants.

Or

"Hydroponics is the art of growing plants without soil, but in water enriched with nutrients and oxygen."

Advantages:

1. It can reduce the physical work involved in growing.
2. It can reduce the amount of water used in growing.
3. It can save on space more than can be grown in the same area.
4. Promote faster growth.
5. No weeds and reduced pests.
6. Good plant quality produce and taste.

Disadvantages:

1. Large system high energy costs can be prohibitive.
2. Moisture from systems can cause damage to buildings. Fish tanks do this also. Can be an advantage too.
3. Specially formulated, soluble nutrients are needed in most systems.
4. Pests and diseases can remain a big risk.

5. Indoor plants can get aphids which transfer easily.
6. Some water born diseases can spread rapidly in recirculation system if not monitored correctly.

Types of Hydroponic Techniques:

1. Passive Wick system
2. Nutrient film technique system or Gutter system
3. Deep water culture system
4. Ebb & Flow or Flood & Drain sub-irrigation
5. Aeroponics
6. Drip (with either a recovery or non-recovery process)
7. Aquaponics system

Wick System

The wick system is passive, with no moving parts, and is also the simplest type of hydroponic system. The nutrient solution is drawn into the growing medium from the reservoir with a wick. The grower using this system can use a variety of growing media such as perlite, vermiculite or coconut fibre. However large plants tend to draw and use the nutrient water at a faster rate than the wick can supply it.

Nutrient Film Technique System

Nutrient film technique (NFT) is a hydroponic technique wherein a very shallow stream of water containing all the dissolved nutrients required for plant growth is re-circulated past the roots of plants in a watertight gully, also known as channels.

NFT systems do not use a growing medium. Instead the plants are supported by a basket and the roots dangle in the nutrient solution. There is a constant flow of nutrient solution which means a timer is not required for the submersible pump. The nutrient solution is pumped into the growing tray using a tube it flows over the roots of the plants, and then drains back into the reservoir. This is an inexpensive method as it does not require the expense of replacing the growing medium for each successive crop. Roots dry out rapidly however during power outages or equipment failure when the supply nutrient solution is interrupted.

Deep Water Culture System

The water culture system is the simplest active system to use. A styrofoam platform floating on the nutrient solution holds the plants. Oxygen is supplied to the roots of the plants through a bubbling air stone that is attached to an air pump. Water culture is the system most often used for leafy vegetables such as lettuce that require fast growth and ample water. It is not suitable for most other plants that require a longer growing period. It is suitable for big size plants, there is better water temperature control and it is usually suitable for home hydroponics.

EBB & FLOW OR FLOOD & DRAIN SUB-IRRIGATION

The ebb and flow is a versatile system that floods the grow tray with the nutrient solution for a short period and then drains the solution back into the reservoir using a submerged pump and timer. The timer cuts in several times a day and as it cuts in the nutrient solution washes onto the tray, then as it cuts out the solution drains back into a reservoir. The frequency is dictated by the type of plants being grown and the growing medium used. Several types of growing media such as perlite, rockwool, gravel or grow rocks can be used in this system. It is advisable to use medium with greater water retention abilities such as rockwool as the incidence of root dehydration during power outages is lessened. In this technique, Soilless media is used in containers; water is not supplied continuously and usually 50% less fertilizer concentration is used.

Aeroponics

The aeroponic system is probably the most high-tech type of hydroponic gardening. Like the previous system, the plants are suspended in the air. The root system is periodically (every few minutes) misted with the nutrient solution using a timer and nutrient pump on a short cycle for a few seconds at a time. Due to root exposure in this system the roots can dry out rapidly during power outages or equipment failure.

Advantages of Aeroponics

- Good supply of oxygen
- Disease free systems
- Good for research on root morphology

Disadvantage

- Requires sophisticated mist systems.

Drip (With Either a Recovery or Non-Recovery Process)

Drip systems are simple to operate, very widely used and the most popular system worldwide. A submersed pump is controlled by a timer: when the timer turns on the nutrient solution drips onto the base of each plant using a drip line.

There are two types of drip systems: recovery and non-recovery systems.

Drip (With Recovery Process)

The recovery system is where the excess (drained off) nutrient solution is recovered in a reservoir and then recycled. This system uses a less expensive (and less precise) timer than the non-recovery system. It is more efficient in that it recycles nutrients and due to this does not need to be as precise as the non-recovery system. However this system needs to be carefully monitored to ensure the nutrient solution doesn't vary too much in pH and strength due to the recycling process.

Drip (With Non-Recovery Process)

The non-recovery system does not collect or recycle the excess nutrient solution. It needs to be more precise than the former system to ensure that the plant gets the correct level of nutrient solution and that the runoff is kept to a minimum. This system does not have the same level of maintenance as the former system as the nutrient solution is not recycled so pH and nutrient levels should always be correct.

Aquaponics System

8. Aquaponics is the method of growing fish and plants in recirculating, symbiotic, constructed system.

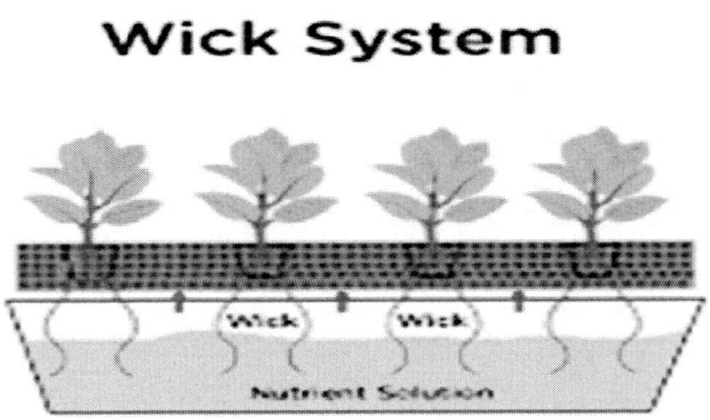

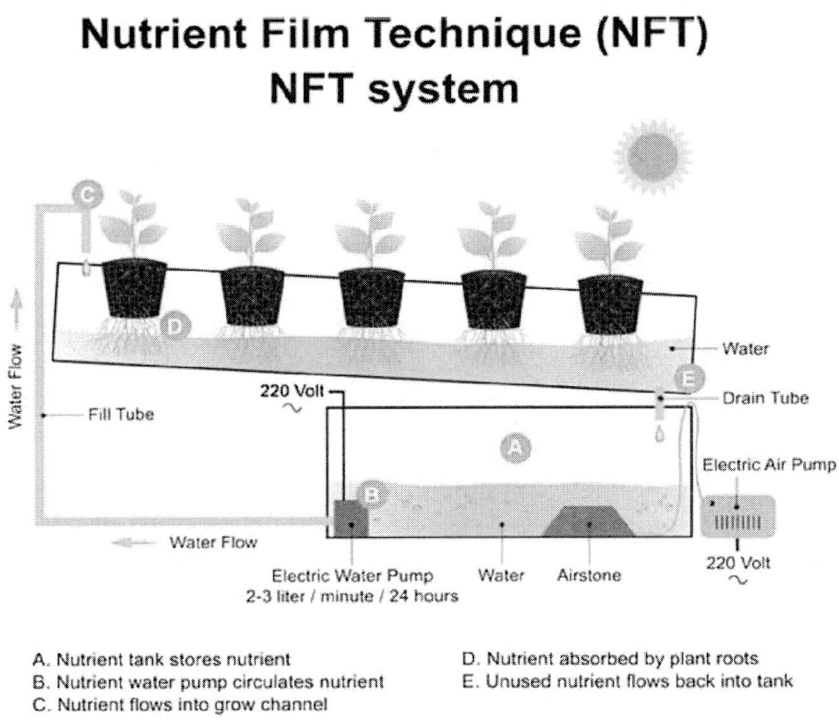

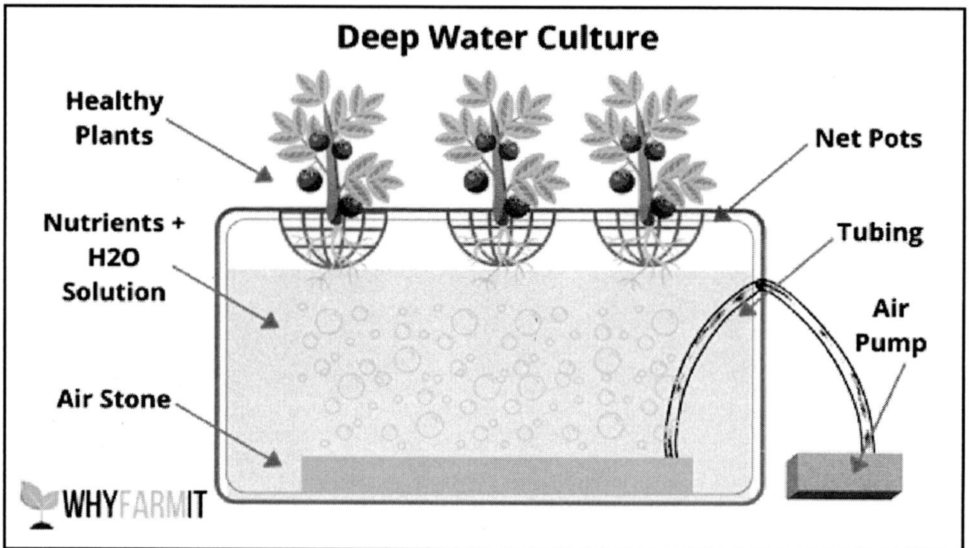
Deep Water Culture
Healthy Plants
Net Pots
Nutrients + H2O Solution
Tubing
Air Pump
Air Stone
WHYFARMIT

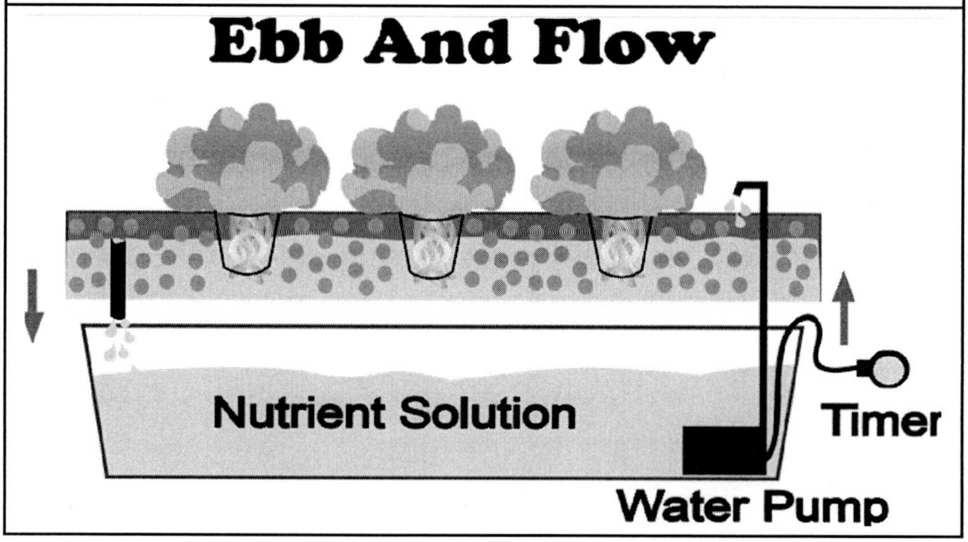
Ebb And Flow
Nutrient Solution
Timer
Water Pump

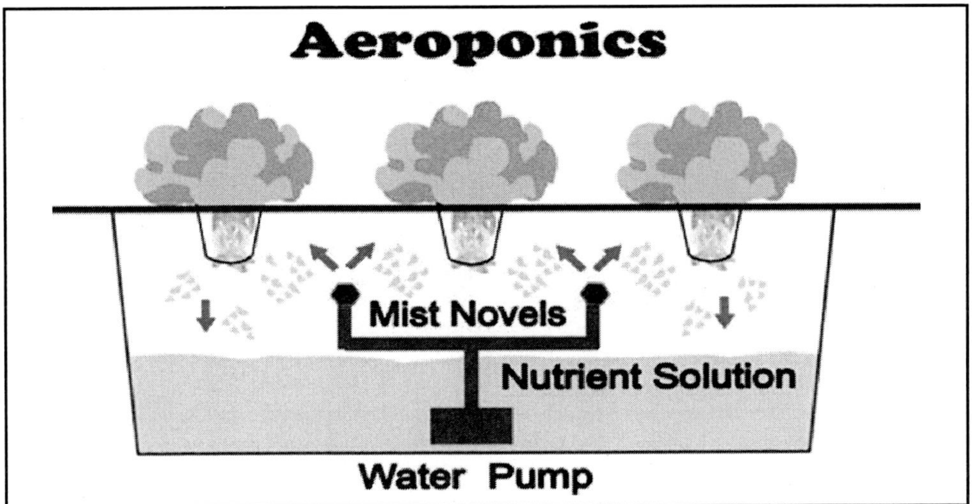
Aeroponics
Mist Novels
Nutrient Solution
Water Pump

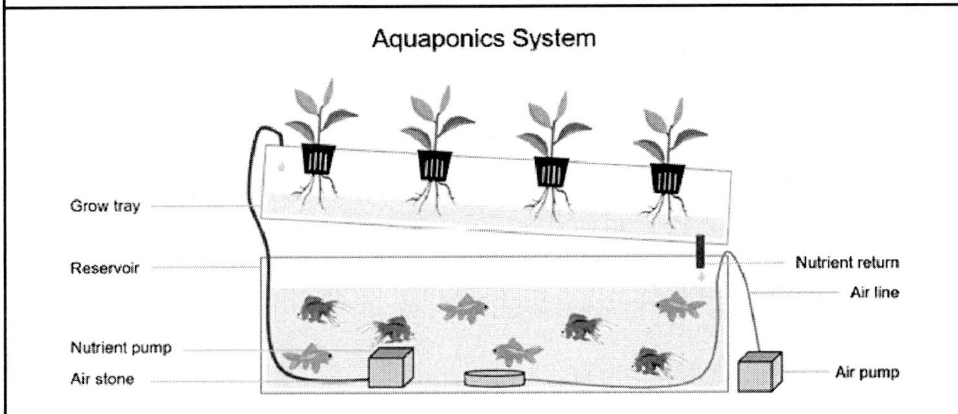
Aquaponics System
Grow tray
Reservoir
Nutrient pump
Air stone
Nutrient return
Air line
Air pump

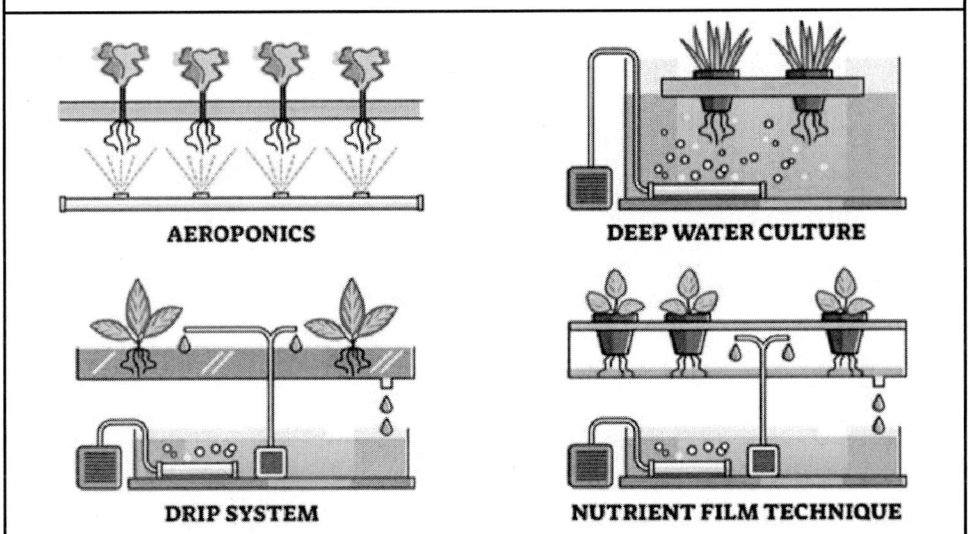
AEROPONICS
DEEP WATER CULTURE
DRIP SYSTEM
NUTRIENT FILM TECHNIQUE

Exercise No. 23

OBJECTIVE – Study on Vertical gardening

VERTICAL GARDENING

Vertical Gardening is a special kind of urban gardening suitable to small spaces, particularly for decorating the walls and roofs in various styles. This is an alternative method for gardening by expanding the scope of growing plants in a vertical space. Intensive urbanization has left hardly any horizontal space for outdoor gardens. Green walls are not only spectacularly beautiful, but also helpful in enlivening the ambiance. Green walls can absorb heated gas in the air, lower both indoor and outdoor temperature, providing a healthier indoor air quality as well as a more beautiful space.

TYPES OF VERTICAL GREENING SYSTEM

1. Green façade
2. Living/green wall

    a. Modular green walls

    b. Vegetated mat wall

Green Facades

1. Green facades are a type of green wall system in which climbing plants or cascading groundcovers are trained to cover specially designed supporting structures.
2. Plants are either grown in the ground or in the elevated containers where they are watered and fertilized.

Green facades are a type of green wall system in which climbing plants or cascading groundcovers are trained to cover specially designed supporting structures. Rooted at the base of these structures, in the ground, in intermediate planters or even on achieving full coverage. Green facades can be anchored to existing walls or built as freestanding structure, such as fences or columns.

Living Walls / Green Wall

Living wall system composed of pre-vegetated panels, vertical modules or planted blankets that are fixed vertically to a structural wall or frame. These panels can be made of plastic, expanded polystyrene, synthetic fabric and support a great diversity of plants species (eg: a lush mixture of ferns, ground covers, perennials and edible plants).

Constructed from pre-vegetated panels, vertical modules or planted blankets (vegetated mat wall) that are fixed to structural framework or to a wall Made from steel framework, plastic, expanded polystyrene and synthetic fabric to support a variety of diversity and density of plant species Tend to require more maintenance such as fertilizer and water than green facade systems that are planted into the ground.

Modular Green Wall

Vertical Garden Modules is made up of recycled poly propylene material. It has attractive look, highly durable in nature and it can be easily installed. It provides instant solution for making garden in your residing place.

Green Wall Cups

- Frame/ Supporting panel
- Easily detachable cups/pots

Vegetated Mat Wall

This system, pioneered by Patrick Blanc, it is composed of two layers of synthetic fabric with pockets filled with the plants and growing media. The fabric walls are supported on a framework and backed by a waterproof membrane against the building wall. Nutrients and water are delivered through an irrigation system at the top of the wall.

BUILDING AND INSTALLATION OF GREEN WALLS

Green wall system vary greatly in their design and construction from DIY projects to modular green wall systems. This heading provides information on

- Structures and components for green wall system
- Suitable Plants
- Growing media
- Irrigation and plant nutrition
- General considerations for green walls

Structures and Components for Green Wall System

1. Front Panel
2. Side Panel
3. Stabilizers

4. Bottom drainage tray
5. Geo textile pouch
6. Hanging hook

		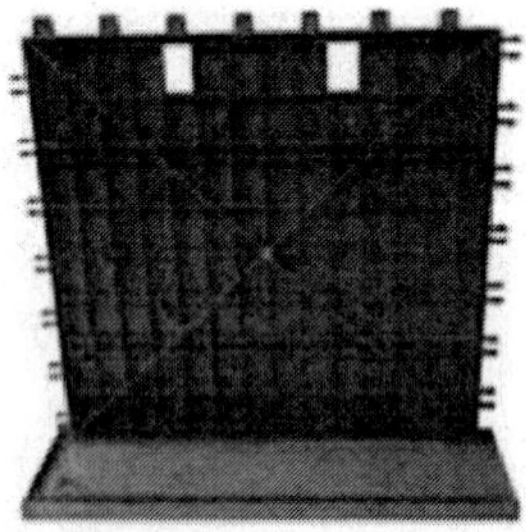
Front Panel	**Side Panel**	**Stabilizer**
		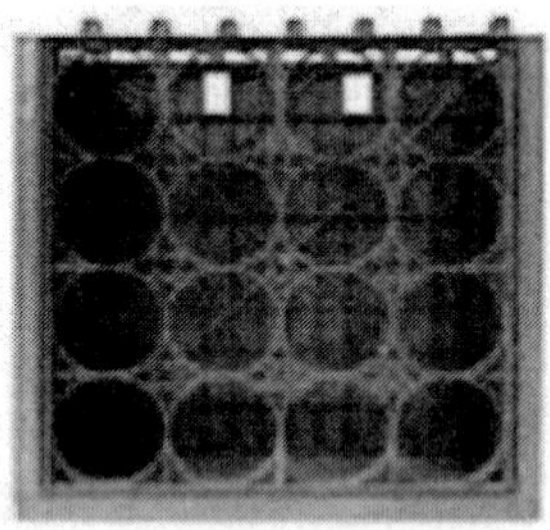
Bottom drainage tray		**Complete Module**

Selection of Plants

The best plants for vertical gardens are dense, compact and low growing. Make sure you choose species that suit the aspect of the wall on which they'll be growing. A wall drenched in sun, for instance, will need a selection of fairly drought-tolerant plants, whereas a shady spot – perhaps under a pergola or verandah – will require a selection of shade lovers, such as ferns. Also bear in mind that the bottom of the unit will hold more moisture than the top, so consider this when you're deciding what to plant where.

Apart from aesthetic preferences and the plants ability to grow in a soilless, vertical location; the selection of plants is based on several factors, for example: What is the local climate like (minimum temperature especially important to know)? How much sun exposure is each area of the surface receiving and how does it change during the year? Is there any particular micro climate — such as high buildings creating strong winds along a wall? Understanding the prevailing growing conditions is essential to make the right decision when choosing plants.

Plants Suitable for Vertical Garden

Outdoor plants like Peperomia, Syngoniums, Philodendron, Epipremnum, Begonia, Anthuriums, Nephrolepis, Chlorophytum, Lantana, Pilea, Rheo discolor, Cuphea, Fittonia, Spathiphylum, Schefflera etc.

Indoor Green Walls/ For Shaded Areas

Herbaceous perennials	*Pepromia, Syngoniums, Philodendron, Epipremnum, Pepromia, Begonia, Anthuriums, Chlorophytum, Pilea, Rheo discolor, Fittonia, Spathiphylum, Schefflera*
Shrubs	*Schefflera, Ficusspp*
succulents	*Rheo discolor, Zebrinapendula, Setcreaseapurpurea*
Ferns	*Nephrolepis*

For Outdoors/Exterior Green Walls

Herbaceous perennials	*Asparagus spp., Pileamicrophylla, Alternenthera, Mentha spp.*
Succulents	*Jade plant, Sedums, Portuluca*
Shrubs	*Dusty miller, Cuphea*
Ground covers	*Baby's tear, Callisarepens*
Grass like foliage forms	*Ophiophogon, Dianellatasmanica*

GROWING MEDIA

Requirements:

1. Weightless media
2. High Water holding capacity
3. High Nutrient holding capacity
4. Good Porosity
5. Neutral pH

Cocopeat, Perlite, Sphagnum moss, vermiculite, vermicompost, shredded bark and leaf molds are the common media combinations used. Soil in not used since it increases the weight of the green walls.

There are three types of growth media used in living walls:

- Loose media
- Mat media
- Structural media

Loose medium walls tend to be "soil-on-a-shelf" or "soil-in-a-bag" type systems. Loose medium systems have their soil packed into a shelf or bag and then are installed onto the wall. These systems require their media to be replaced at least once a year on exteriors and approximately every two years on interiors. Loose soil systems are not well suited for areas with any seismic activity. Reparations are only achieved by re-stuffing soil into the holes on the wall, which is both difficult and messy. Loose-soil systems should not be used in areas where there will be a lot of public interaction as they are quite messy and lose their soil little by little over time. Loose-soil systems with physical media erosion systems are well suited for all green wall applications.

Mat type systems tend to be either coir fibre or felt mats. Mat media are quite thin, even in multiple layers, and as such cannot support vibrant root systems of mature plants for more than three to five years before the roots overtake the mat and water is not able to adequately wick through the mats. The method of reparation of these systems is to replace large sections of the system at a time by cutting the mat out of the wall and replacing it with new mat. This process compromises the root structures of the neighbouring plants on the wall and often kills many surrounding plants in the reparation process.

Structural media are growth medium «blocks» that are not loose, nor mats, but incorporate the best features of both into a block that can be manufactured into various sizes, shapes and thicknesses. These media have the advantage that they do not break down for 10 to 15 years, can be made to have a higher or lower water holding capacity depending on the plant selection for the wall, can have their pH and EC›s customized to suit the plants, and are easily handled for maintenance and replacements. They are the most robust option for a living wall for both exterior applications and for interior applications. They are also the best choice in areas where high-winds, seismic activity or heights need to be addressed in the design. Structural media are superior to the other media for their longevity and high-level of performance in a variety of circumstances. Depending on the installation, they do tend to be more expensive to install, but lower cost to maintain.

IRRIGATION AND PLANT NUTRITION

Green walls cannot be sustained without irrigation. Interruptions to the water supply are a common cause of plant failure on green walls. Systems designed with inbuilt irrigation should mitigate plant losses due to inconsistent moisture management, although errors can still occur. Automated, remotely controllable irrigation systems are used for walls in high profile locations, or in situations where access is challenging. Note that the quality, design and costs will vary between different systems. The most sophisticated systems enable the maintenance supervisor to keep track of the automated performance of the system, including the volume of irrigation delivered, its frequency, substrate moisture content, as well as pH and nutrient levels in the water supply. The settings can be overridden if needed; for instance, the frequency or duration of irrigation cycles may be increased on hot days. In hydroponic systems, plant nutrition is delivered by a fertilizer injection system that releases controlled doses of fertiliser into the irrigation system (fertigation). Management of fertigation systems and rates of delivery requires specialist knowledge, as it is more complex than fertilizing soil or growing media. Hydroponic systems require continual monitoring of pH, water hardness and total dissolved solids (TDS), and adjustment of these parameters where necessary.

Irrigation

The irrigation system is designed to minimize water consumption. It consists of an automation-unit with equipment for control of nutrient injection and irrigation cycles. When a surface has a variation of sun exposures, the irrigation is divided into segments in order to program it specifically for each part. Within the multi-layered felt surface a drip-tube is integrated. Water consumption varies with heat and sun exposure, but compared to normal green spaces or a lawn, the consumption is normally lower. It averages between 2-5 l/m2/day.

For hydroponic green wall systems, the fertigation system may apply 0.5-20 litres of irrigation solution per square metre perday. Internal green wall requirements are at the lower end of this range, and external green walls at the higher end. Irrigation cycles typically last a few minutes and will be required several times a day. Keeping irrigation volumes low minimises waste and reduces run-off. Irrigation run-off may be captured in a tank at the base of the wall and recycled back through the green wall system. Green walls that use a high quality, water-retentive growing medium, and are not in an exposed or particularly hot location, may thrive on a weekly watering regime. In most simple, soil-based systems, including DIY systems, controlled release fertiliser is mixed in with the growing medium, rather than using a fertigation system. Irrigation must be available as soon as the plants are installed in the wall system. The irrigation system requires a water meter to monitor irrigation volume, and a pressure gauge to monitor the even application of water. The need for ongoing regular irrigation and the expectation that water will be used sustainably means that stored (harvested or recycled) water should be used whenever possible, so a pump is necessary.

Design / Process

Each vertical garden is given a unique design and selection of species. The composition of plants takes in consideration the specific environment where it will be built, such as the local- and micro climate, sun exposure and the surrounding context. The aim is to create a one of a kind and site-specific garden that stands beautiful through all the seasons of the year.

A well executed design is also a way to minimize the future maintenance demand of the garden. A plant's growth habit, size and behavior on a vertical surface is important knowledge for making the right combination of species, in order to keep the competition between plants at a healthy level. Choosing the right plant for the right place makes sense for any garden, but maybe even more so in a vertical garden.

A vertical garden can be installed in almost any location and as a living material, the potential of integrating plants in our urban environments is interesting. Places never thought of as possible could be inhabited by plants, like subway stations or other intensely frequented places where horizontal space is difficult to spare.

Light

Direct sunlight can deliver over 100.000 lux whereas the average light level in an office is around 300-500 lux. Even if the least light demanding species are used, artificial light is normally necessary indoor. A few species will stay fine at 900 lux, but a slightly increased level at some parts of the surface will broaden the variation of species that can be used. An artificially illuminated surface has shifting light levels, due to the fact that light reduces with the square of the distance from the light source. Some areas might have 3.000 lux and others 900 lux. The plant design is made with this in mind, taking advantage of the higher levels for more demanding and interesting species.

General Considerations

- Watering: Appropriate time
- Carefull selection for wind prone areas (Succulent and hardy plants)
- Removing the dried leaves
- Keeping the structure clean
- Disposing the water from drainage system
- Pruning if necessary
- Timely application of fertilizers

Benefits of Vertical Gardening

- Aesthetic effects
- Acts as natural insulation for hot and cold air and a save energy for your building
- Reduces CO_2 levels and increases oxygen and improved air quality
- Conserves water and watering takes less effort
- Sound absorption and noise absorption
- Improves thermal insulation and energy efficiency
- Provides protection to buildings from adverse temperature and hence improves the life expectancy of the buildings
- Mitigate urban island heat effect
- It holds rain water, providing food and shelter for wildlife

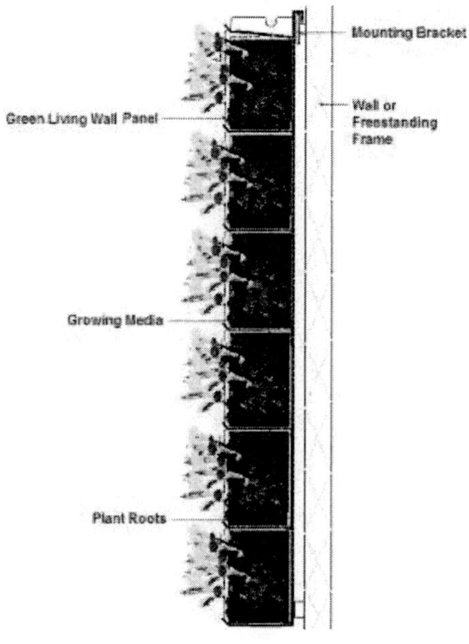

Vertical Garden

Indoor Vertical Garden	**Public Building/Apartments**

Malls Vertical Garden	**Universities/Hospitals/ Airports**

Terminology

Adventitious bud: The bud arising from places other than the leaf axile and shoot terminals.

Adventitious root: The root arising from the stem or shoot or leaves.

Aggregate fruit: It is a type of fruit develop from a single flower having apocarpus pistils. The carpels being free develop into fruits/fruitlets and the aggregation of entire fruits/fruitlets give rise to a single fruit. For example raspberry, strawberry, blackberry etc. Simple and succulent/fleshy fruits are developed from syncarpus pistils/fused carpels

Allogamy: The fertilization of a flower by pollen from another flower, especially one on a different plant. In other it is called cross-pollination.

Amphimixis: Multiplication of plants through seeds eg papaya, jamun, phalsa and mangosteen.

Andromonoecious: It is type of the pistillate flower of monoecious is replaced by bisexual flower as in muskmelon.

Anthesis: It refers to opening of flower bud and time of fertilization.

Apomixis: Asexual reproduction through seeds (cloning of seedlings per seed). Recurrent apomixes (*Malus* spp., *Rubus* spp. Non-recurrent apomixis (Lilium spp. *Solanumnigrum*), Nucellarembroyony or adventitious embryony (Citrus), Vegetative apomixes (Garlic, Agave, *Dioscoreabulbifera*).

Apomixis: Asexual development of embryo in the ovary without fertilization.

Apospory: A form of apomixes in which the embryo sac develops from a vegetative cell of the ovule.

Arboriculture: Growing purpose of trees for aesthetic/scientific/educational

Aril: Outer growth of seed by expansion of the stalk of the ovule after fertilization eg. pomegranate, litchi.

Autonomic parthenocarpy or vegetative parthenocarpy: A type of parthenocarpy in which fruits develop without pollination and fertilization.

Autopolypoid: A polyploidy with more than two copies of the same genome. Eg. three sets of same genome called auto triploid (banana)

Balausta: The epicarp is leathery and is crowned by sepals. The seeds remain covered with juicy layer of the pulp which is developed from seed coat. For example pomegranate.

Balausta: A type of fruit which is modified form of berry, inferior, many celled and many seeded with leathery pericarp. eganar.

Bark: The outer tissues of the cambium layer of woody stem or roots.

Berry: Many seeds remain enclosed in pericarp of the fruits. Eg.guava, papaya, tomato banana etc.

Berry: A simple fruit derived from one flower in which a parts remain succulent. The true berry is derived from an ovary (grape) the false berry from an ovary plus receptacle tissue (blue berry).

Biennial: Plant that completes its life cycle in two years or growing season, with a dormant period in between. It produces leaves the first year and flowers and seeds in second.

Blanching: Protecting plant stems (leeks), bulbs (fennel) or head (cauliflower) from sunlight and the elements. This enhances succulence and prevents discoloration.

Blight: A disease characterized by sudden, severe, and extensive spotting, discouration, wilting or destruction of leaves, flowers, stems, or entire plants, usually attacking young, growing tissues (in disease name, often coupled with the name of the affected part of the host, e.g. leaf blight, blossom blight, shoot blight)

Bottom heat: A technique for rooting of cuttings which entails the application of artificial heat to the rooting medium to keep the basal end of the cutting comparatively warm for inducing rooting.

Bridge grafting: A grafting method in the injured plants. Where one year old scion inserted to save the tree. The scion usually collected during the dormant season and are inserted in the tree when the bark begins to slip.

Bud: It is region of meristematic tissues generally protected by modified scale leaves and has the potential for developing into leaves, shoot and flower or a simple lateral leaf bud which is placed on the stock for further growth and development into, leaf, shoot and flowers.

Bud break: Stage of bud growth after rest or sprouting of dormant bud as a response to improved growth conditions.

Bud nicking: A deep knife incision through the vascular stands into the wood directly below a bud to retard its growth.

Bud notching: Removal of a wedge shaped piece of bark with wood directly above a bud to stimulate growth.

Bud sport: A new strain or clone arising from a mutated bud.

Bud stick: A shoot or small branch used as a bud source during grafting.

Bulb: Specialized underground organ consisting of a shoot,shortened fleshy vertical stem axis, apex, flower primordia and are covered by thick fleshy scales eg. onion

Bulblets: It is secondary bulb forms in that forms in the angle between a leaf and stem or in place of flowers on certainplants. Eg. onion, garlic etc.

Callus: The undifferentiated mass of parenchyma tissue at the edges of a bud, graft or in vitro cultured explants.

Cambium: A thin layer of tissues located between the bark (phloem) and wood (xylem) of vascular plants.

Catkin: A scaly cylindrical floral spike.

Chilling requirement: A period required by certain plants and plant parts in order to break physiological dormancy or rest. The chilling requirement is expressed in terms of the required number of hours at 7 C or less.

Chimera: A plant or plant parts composed of genetically different layers. Eg. crotons, duranta etc.

Chimeras: A plant consisting of cells of two of more genotype in the same part of a plant. It may result from mutation, irregular mitosis and somatic crossing over.

Climactric: Fruits experiencing sudden upsurge in rate of respiration at the time of ripening. Eg. Mango, Guava, Papaya, Jackfruit, Fig, Sapota, Passionfruit.

Clone: A group of plants originated as asexual progeny of a single mother plant.

Cloning: Multiplication of a clone by means of vegetative propagation.

Cold frame: An enclosed (glass or plastic) unheated frame used as a propagation unit and for growing or protecting young plants in early spring in temperate climate.

Cole crops: The Word cole has come from latin 'caulis' meaning shortened stem. Thus crops where the edible portion is stem and leaves are called cole vegetables. These are generally cool season vegetables and come from families mainly cruciferae and composite. They grow in a temperature range of 15 C-20oC and pH of 6.0-6.5. Common vegetables coming under this group are

- Heading type: cabbage, Brussels sprout and lettuce
- Curding types: cauliflower and sprouting broccoli
- Leafy types: Kale and Savoy cabbage

Condiments: are also spices, products of which are used as food adjuncts to add taste only example vanilla, saffron, asafetida, garlic etc.

Cool season vegetable: They are generally growing at a temperature of 15 C to 25 C and are generally crops in which the edible portion is root, stem, leaves flower and buds as in carrot, cauliflower, lettuce and globe artichoke except a few like peas where fruit is the edible part.

Corm: It is a solid and swollen stem with distinct nodes and internodes enclosed by the dry, scale like leaves eg. Yam, Elephant foot yam etc.

Cormel: Tiny corms produced around the base of the mother corm eg. Colocasia

Crown grafting: A method of grafting in which a mature dormant wood of the previous season growth is grafted on the stock in the crown region.

Cryopreservation: Storage of seed material in liquid nitrogen N_2 @ -196oC and liquid carbon dioxide -43 C.

Cultivars: It is an assemblage of cultivated plant which is clearly distinguished by any characters (morphological, physiological, chemical or other) and which when reproduced sexually or asexually retains its distinguished characteristics. The word cultivars is contraction of the words "cultivated variety"

Curd: white Inflorescence of cauliflower

Cutting: Cutting may be described as a method of propagation where detached vegetative plant part and placed under favourable conditions with the objective to develop successful individual plants.

Cybrid: Hybrid cytoplasm resulting from protoplast fusion.

Damping off: A complex of water molds (fungi) that attack seedlings, causing them to wilt and die. These fungal pathogens- *Phtopthora, Pythium* and *Rhizoctonia*- thrive in wet, humid conditions and are the causal agents.

Day neutral plant: A plant that will flower under any day length eg Tomato, chilli okra etc

Dehiscent fruits: In such fruits pericarp splits and open at maturity and the seeds are released. Eg. pod, capsule, silique, silicula, nutetc

Diara cultivation: Riverbed cultivation. Growing of cucurbits in river beds or river basins constitute a distinct type of farming. These arecalled "diara lands" in UP and Bihar.

Dioecious: where staminate and pistillate are produced in differentplants e.g. pointed gourd, papaya etc

Distal: The distal end of either the root or the shoot is that furthest from the stem-root junction of the plant and nearest to the tip of the shoot or root.

Dormancy: The condition of a bud or seed characterized by lack of visible growth.

Double working: The practice of inserting the mutually compatible inter- stock piece between stock and scion to more than two kinds of plants together by grafting.

Drenching: Process of applying diluted solution of agro chemical or any other substances directly to the base / root zone of theplants for effective treatment or deep penetration.

Duodichogamy: A condition when two batches of male flowers are temporarily separated by a batch of female flower in between, the situation is known as duodichogamy. Theresting period between flowering avoid selfing completely.

Earthing up: It is the techniques of pilling soil up around the base of a plant with a objective to provide strength or cover roots or improving rhizosphere. It can be done by hand or withpowered machinery.

Epicotyl grafting: (stone grafting): Grafting in 7-10 days old seedling stocks are grafted with current year's scion shoots by both splices and cleft methods, practiced in mango, cashew nut etc.

Etiolation: Development of plant parts or plants in absence of light, accumulation of auxin helps root formation.

Explant: Living tissue use in tissue culture for the propagation may be the part of lead, stem, root, bud, anther etc.

Fertigation: Fertilization with irrigation water.

Floating Vegetable gardens: This is a very artistic and peculiar type of vegetable gardening. It is practiced in Dal Lake in Kashmir. Most of the vegetables supplied to Srinagar are produced in floating garden. To grow vegetables, a floating base is prepared from the roots of typha grass. This grass grows naturally in wild form in the lake. When floating base is ready, seed sowing or transplanting is done over leaf compost available locally in lake itself. Time to time as and when needed, intercultural operations like removal of weeds, watering, application of fertilizer, pesticides etc. are made through boats.

Floriculture: It is another branch of horticulture which deals on commercial floriculture, landscaping and cut flowers.

Formal garden: First the plan is made on the paper and then land is selected accordingly. Plan is symmetrical. These types of gardens are of geometric design. (square or rectangular). Therefore the roads are cut at right angles. Flowers are also of geometric shapes. The arrangements of trees and shrubs are necessarily geometrical and kept in shape by trimming and training. Other features like fountains, Pool, cascade etc are used for further attraction. Eg. Persian gardens and Moghul gardens.

Fumigation: method in which materials are treated in close chamber byfumigant to kill the storage pest.

Gietonogamy: It is a greek word derived from geiton-neighbour and gamein- to marry. Thus, it is the process of pollination of flower by another flower of same flowering plant. The process interferes with sex function and leads to reducedoutcrossing. For example- orchids, black pepper

Graft incompatibility: Inability of stock and scion to form successful graft union.

Graft union: The site of grafted plant where the scion and rootstock are united.

Grafting: It is an art of inserting a part of one plant into another plant by exposing the actively growing tissue, so that they will unite and continue their growth as one plant.

Grafting wax: It is a material applied on the graft union in order to minimize desiccation and exclude water access.

Gynoecy: The condition of species of which only the females areknown. Eg. bittergourd

Gynomonoecious: It is the condition where pistillate and perfect flower areproduced separately eg. cucumber

Habituation: A condition in which callus and cell systems lose their requirements for specific growth regulator and continue to grow. This trait develops in tissue cultured plants after long exposure to high concentrations of a particular growth regulator. Sometimes, shoot culture becomes habituated to cytokinin and continues to proliferate even after shifting to medium without growth regulator.

Haploid: Cell contains a single set of chromosome.

Hardening off: The process of gradually acclimatizing a plant that has been raised indoors to lower temperatures or more severe conditions, so that it is not severely checked when plantedout.

Head: edible part of cabbage

Hedge fencing: A live fence around the orchard use with the objective to protect orchard from stray & wild animals along with human interference. *E.g.*Euphornia, Opuntia, Karonda, *Inga dulcis* etc.

Hermaphrodite: It is the flower bearing both male and female organs.

Hesperidium: Type of fruit in citrus, a modified berry many celled fleshy fruit which develops from a polycarpellary superior ovary with axile placenta ion. Such fruit epicarp and mesocarp fused together forming rind, endocarp remain thin and papery and inner part having juicy succulent hairs.

Heterodichogamy: It is state of having genetically reciprocal morphs which function as a female, first then as a male, the other behaving in opposite ways. It is observed in walnut, hazelnut, *Annonasquamosa, Zizyphus jujube,Cinnamomumcommiphora*etc)

Hi tech poly house: Hi tech poly house is high cost protected structure in whichenvironmental factors can be controlled as per desire by various automated provisions of heating and cooling.

Homogamy: In this condition, the stamens (male=androecium) and the carpels or pistils (female = gynoecium) of the flowers mature at the same time. This process enhances the chance of pollination. Homogamous condition is seen in apricot,citrus, peach and dwarf coconut cultivars.

Horticulture: It is defined as the crop science which deals with the production, utilization and improvement of fruits, vegetables, ornamental plants, spices, medicinal, aromaticand plantation crops.

Hyperhydricity: Also termed as vitrification it is expressed by translucent, water soaked and succulent appearance of the culture as a result it fails to proliferate. It is accounted by excess uptake of water and inhibition of lignin and cellulose synthesis.

In vitro: Latin word meaning within glass.

In vivo: Latin word meaning with in the living organism.

Indexing: It is a process to test mother plant for freedom of virus diseases by budding a scion on susceptible plants.

Informal garden: This style reflects naturalistic effect of total view and represents natural beauty. In this style, plan is asymmetrical and according ao the land available for making garden. Roads and paths are made with curvatives and bending. Water bodies are made of irregular shapes. Hillocks are made to create natural mountainous scenery. Flower beds are made in natural form and instead of trimming, annual pruning is done. Eg. Japanese garden.

Integrated nutrient management: It is an integrated approach which utilizes all the organic, inorganic and biological components to maintain the soil fertility as well as ensured nutrient supply to plant forsustainable production.

Integrated pest management: It is pest management program that integrates regular pest control practices and scientific principles for a more environment friendly approach. It ensures the minimum use of pesticides and utilizing the cultural, biological andstructural strategies to control pest problems.

Inter-node: The region of a stem between two successive nodes.

Isolation distance: It is the minimum separation required between two or more varieties of the same species for the purpose of keeping seed pure.eg Cole crops (1600m foundation seed, 1000 certified seeds), Onion (1000m foundation seed, 500m certified seeds), ginger and turmeric (10m foundation seed, 5m certified seeds), legumes (50m foundation seed, 25mcertified seeds) etc.

Juvenile stage: Early or vegetative phase of growth characterized by carbohydrate utilization.

Kitchen garden: It is established with the less available area, choice of varieties, not restrictly followed for high yielding varieties. It is both intensive and intercropping no need of market. In other words, it is laid out in the backyards utilizing the available space and time, growing the seasonal vegetables and a few perennial vegetables and fruit trees to providefresh vegetables for the domestic use.

Lanoline: The natural fat of wool generally used as a carrier of growth substances.

Lath house: A propagation structure made up bamboo strips which provide out door shade and protectd containers grown plants from high summer temperature & intensity.

Layering: It is a vegetative propagation method in which the developments of roots on a stem while it is still attached to the mother plant or parent plant.

Leaf vegetables: It includes Amaranthus, Spinach, palak, fenugreek, NewZealand Spinach, coriander where economic part is leaf.

Long day plant: These are the plants which require long light period with short dark period (8-10 hours) to produce flowers eg Potatocabbage cauliflower, knolkhol radish, carrot etc

Male sterility: Non-functional gametes due to chromosome aberration e.gonion, cucurbits

Manures: It is the decomposed form of dead plants and animals,which is applied to the soil to increase production. It is a natural form of fertilizer and is cost-effective.

Market garden: Develop with the urbanization. Practiced in suburbans of cities. Usually farm size is small. Too intensive cultivation of vegetable is practiced. Early maturing as well as off- season vegetables receives priority. Commonly, sale isarranged by the growers.

Matrix: A portion of scion remains inserted in rootstock called matrix.

Maturity indices: Maturity indices are those parameters which indicate that the produce is ready to harvest. At this stage, the crop hasattained maximum growth and maturation.

Medicinal Plant: Whole plant or individual plant part such as root, stem, leaves, bark, flower, fruits, seeds etc or the chemicals derived from these parts are used in different system of medicines (Allelopathy, Ayurveda, Homopathy, Unani etc.) to cure disease are known as medicinal plants. The utilization of medicinal plants are direct utilization in the form of plant part (root, stem, leaf, seeds and barks), powder, extracts, medicinal chemical (alkaloids, glycosides) and plant drugs.

Minispinkler: Used for irrigation of seasonal crops like vegetables, onions, potato, nurseries etc. Made of engineering plastic and available in different connection options. Compact design for easy installation, riser can be used to increase installation, riser can be used to increase the height upto150 mm (6").

Mist propagation: Vegetative plant propagation inside a chamber under spraying water in mist form to reduce water loss through transpiration from the propagating materials.

Monoecious: It is an advance from where both staminate and pistillate flowers are produced separately in same plant e.g.cucumber, bottlegourd, bittergourd, watermelon.

Monoecy: When male and female flowers are present on separate branches of the same plant, this situation is known as monoecy and the flowers are known as monoceious. It is found in banana, jackfruit, coconut, peacanut, hazelnut,cucurbits etc.

Mulching: A material applied to the surface of a soil to conserve moisture, stabilize soil temperature, suppress weed growth, protect plant roots from heat or cold or to keep fruit clean Multiple/composite

Fruit: This is a type of fruit which develops collectively fromentire flowers of the whole inflorescence.

Necrosis: Actively growing tip sometimes develop tip die-back and it dries. This is due to calcium deficiency and its alteration requires refining basic medium.

Non-climacteric: Fruits experiencing simple gradual decline in rate of respiration at the time of ripening. Eg. Litchi, Lemon, Lime, Orange, Grape Pomegranate, Pineapple

Nursery: It is place where plants are reared until they are ready until they are ready for final transplanting in the field. It is of two types: temporary and permanent. Temporary type of nursery is raised in open especially under sheltered position of tree or even in totally open condition. For such nurseries no provision of permanently walled bed is made. Permanent type nurseries usually have permanently walled- bed. Permanent type nurseries are generally permanently walled and often provided with overhead covering against rain or frost. There may be facility of overhead irrigation system. In permanent nurseries there may be permanent provision of disinfecting or sterilizing the soil. Making facilities for such arrangement makes the raising of nursery costly and hence permanent nurseries are used mostly forcostly, highly demanded and hybrid seeds.

Oleoresin: are semi-solids extracts composed of resin and essential or fatty oil, obtained by evaporation of solvent used for their production eg pepper 35-60% piperine, ginger (25-30% gingerene), Chilli (2-4% capsaicin), Turmeric (20-30%cucumin), coraindar (1.5-2% D-linalool).

Olericulture: It is the branch of horticulture which deals on Vegetableslike leafy vegetables, root, tuber, cole crops etc

Organic: Of plant or animal origin, containing carbon compounds.

Ortet: The original plant from which a clone has been derived.

Orthodox seeds: Desiccation tolerance during development and may be stored in the dry state for predictable periods under definedconditions.

Orthodox seeds: are those seeds that could be dried to low moisture content and tolerate freezing temperatures eg Guava, Sapota,Apple, Cherry etc.

Parthenocarpy: Development of fruit without pollination and fertilization.

Parthenogenesis: Fruits develop without pollination and fertilization but stillthey produce viable seeds e.g. Mangosteen, strawberry

Parthenogenesis: Development of an egg without fertilization.

Pedicle: The stalk bearing flower of an inflorescence.

Peduncle: The stalk supporting either an entire inflorescence or a solitary flower.

Pepo: It is a modified berry in which epicarp becomes hard andtough. The seeds remain attached to the plancental hair of endocarp. For example citrus fruits

Perennial vegetables: are vegetables that can live for more than two years eg.Artichoke, Asparagus, Drumstick, Rhubarb, curry leaf, chekkurmanis, chow-chow etc.

Periphysis: Variation of growth habit of vegetative propagules due to their origin from different environments, such as shade and sun leaves on an individual tree.

Pin type: In this type, style is longer than stamens. It is broad toward apex and narrower towards base. Plants with pin type of heterostyly set more seeds as they have exposed stigmas. Itis found in pomegranate and sapota.

Plant growth regulators: These are either synthetic compounds or plant hormones that modify plant physiological processes at very lowconcentration.

Plant hormones: These are the organic substances other than nutrients which are produced by plants away from site of action and regulate plant's physiological processes at very lowconcentrations.

Plantation crops: Grown extensiveareaanduseafterprocessing eg.Arecanut, Tea, Coffee, Rubber,

Plug trays: Plastic made up trays having numerous cavities or small compartment in which seeds are sown and seedlings areallowed to grow in controlled conditions.

Polarity: A state of having two opposite poles of cuttings should be planted in that position and fashion where it was already attached to the mother plant i.e., to to bottom with the gravity for good sap flow in it and success.

Pollination: The transfer of pollen from the anther to stigma.

Pollinator: An insect or other source by which pollen is carried from one flower to another.

Pollinizer: The producer of pollen or the cultivar used as a source ofpollen for cross-pollination.

Poly embryony: A condition where one seed produce 2 or more embryos or seedling. e.g. in mango, citrus, apple.

Polyembryony: Formation of multiple embryos in a single seeds (more than one seedlings per seed). Eg. Mango, citrus, jamun.

Polytunnel: A structure covered with plastic films used for plant protection. Walk-in-polyhouse (high tunnels) are approximately 6 to 8 ft. high at the ridge, low tunnels arerarely more than 18-ft high

Pome: It is two or more chambered false fruits which are developed from fleshy and edible thalamus. Four or five carpels are united to a fleshy receptacle to form the fruit. The papery and leathery endocarp contains one or more seeds. The fruit develops from syncarpus inferiorgynoecium e.g. apple, pear, quince.

Pomology: It is the branch of horticulture which deals on scientific cultivation of fruit crops like crops like Litchi, Mango Sapota Loquat Papaya

Propagule: A plant part such as a bud, root, shoot etc used to propagate an individual by vegetative means.

Protandry: It is the condition of maturation of anther before pistil e.g.carrot, onion

Protandry: When the stamens mature first than the carpel, it represents protandry condition. It is found in walnut, coconut, sapota,passionfruit, soursop (*Annonamuricata*) etc.

Protected cultivation: It is also called as green house cultivation. It is process of growing crops in a controlled environment. This means that the temperature, humidity, light and such other factors can be regulated as per requirement of the crop. This assists in ahealthier and a larger produce.

Protogyny: When the carpels mature first than the stamens, it is referred to protogyny condition. For example- Annona sp. Except *A. muricata*, fig, banana, plum, pomegranate,avocado etc.

Proximal: The end of either the root or the shoot that is nearest the stem root junction of the plant.

Pruning: It is removal of a portion of a tree to correct or maintain tree structure. It main objectives are regulation of shape and growth of tree. Enhance the production and quality of fruits.

Pruning: Pruning techniques in which three branches are trained and removal of other branches with an objective to get more yield and maintaining ratio of vegetative and reproductivegrowth.

Pureline selection: is a method in which new variety is developed by selection of single best plant progeny among traditional varieties or land races.

Ramet: The individual member of a clone or progeny of a clone.

Recalcitrant seeds: Do not survive drying to any large degree and are thus not amenable to long term storage.

Recalcitrant seeds: could not dried below a relatively critical moisture content and could not tolerate freezing temperature. Desiccation sensitive *e.g.*Jamun, Jackfruit, Mango, Litchi, Mangosteen,Durian Avocado, Citrus, Rabutan etc.

Rhizome: It is a specialized undergroundstemwhichgrows horizontally at or just below the ground surface as in gingerand turmeric

Root crops: Here the edible portion and economic is root, which develops below ground as a result of swelling of the taproot and subsequent accumulation of reserve materials. Root crops are mostly propagated by seeds. Eg. carrot, radish, turnip

Rouging: It refers to the removal of off-type plants, plants of othercultivars and weeds from seed field.

Scarification: Physical or chemical abrasion of seed coat. Mechanical scarification (Ber H_2SO_4 at 6 minutes, peach, walnut), Hot water scarification (guava@ 77 to 100 C soaking in cool water for 12-14 hrs), acid scarification (guava @ 3 min, ber@ 5-6 hrs)

Scarification: Any process of breaking, scratching, mechanically altering or softening the seed cover to make them permeable to water and gases.

Schizocarpic fruit: It is developed from flowers having two or more united carpels which split into indehiscent one seeded parts calledmericarp. Coriander, carrot.

Seed disinfection: Refers to the eradication of fungal spores that have become established within the seed coat, or in more deep-seated tissues. For effective control, the fungicidal treatment must actually penetrate the seed in order to kill the fungus that is present.

Seed disinfestation: Refers to the destruction of surface-borne organisms that have contaminated the seed surface but not infected the seed surface. Chemical dips, soaks, fungicides applied as dust, slurry or liquid have been found successful.

Seed dormancy: It is a physical or physiological condition of a viable seed which prevents germination even in the presence of favorable conditions for germination

Seed dressing: It is the most common method of seed treatment. The seed is dressed with a dry formulation or wet treated with a slurry or liquid formulation.

Seed Priming: Priming is a process in which seeds are imbibed either in water or osmotic solution or a combination of solid matrix carrier and water in specific proportions followed by drying before radicle emergence.

Seed Protection: The purpose of seed protection is to protect the seed and young seedlings from soil organisms which might otherwise cause decay of the seed before germination.

Self incompatibility: Pollen and ovules are functional but fertilization will not take place due to physiological imbalance i.e failure of pollen tube to enter into the egg cell. Eg certain Tomatospecies

Short day plants: These are the plants which require short light period and long dark periods (10-24 hours) to produce flowers eg.clusterbean, cucurbits etc.

Soilless mix: Potting medium that contain a mixture of peat, vermiculite, perlite, compost or other materials but no mineral soil.

Solarization: The practice of heating soil to levels lethal to pests and pathogens through application of clear plastic to the soilsurface for 4 to 6 weeks during sunny, warm weather.

Sorosis: It is developed from spadix or spike. The small fruits fuse by their succulents calyx. The mother axis becomes fleshy.For example, pineapple, jackfruit, mulberry.

Sorosis: A multiple fruit developing from a spike or spadix where the flower fuse together by their succulent sepals and at the same time the axis bearing them grows and becomes fleshy or woody and as a result the whole inflorescence forms a compact mass.

Spadix: A spike with a fleshy axis which is enclosed by one or large, often brightly coloured bracts called spathes.

Spices: are those plants products of which are made use as food adjuncts to add aroma and favoureg. fennel, coriander, cumin etc.

Stratification: Methods of breaking seed dormancy in which the imbibed seeds are subjected to a period of chilling to after ripen the embryo.

Sub-tropical fruit: The fruit plants are grown in region where dry hot summer and less winter. This region receives less rain eg. citrus, phalsa, fig, guava, pomegranate

Syconus: This type of fruit develops fro a hollow, fleshy andenlarged receptacle which encloses numerous male and female flowers eg. fig, gular, banyan.

Take: The slang term for the phenomenon of the successful union of stock and scion and growth of the grafted plant.

Temperate fruits: The fruit plants require chilling hours to break dormancy, thereafter plants reaume growth and have flowering eg. apple, pear, peach, plum, apricot,walnutetc

Thermosensitive/ vernalization: Low temperature treatment required for cole crops and root crops to transform from vegetative to reproductive stage. For eg. European carrot requires 7oC for 2 months forbolting.

Thinning: Pulling or clipping the weak seedlings in a pot or row in order to leave the others room enough develop. Removing a branch or water sprout at the point where it joins a main stem, branch, or trunk.

Thrum type: Such plants possess flower with a short style. The stamen is longer than style. Thrum type heterostyly flower with a short style. The stamen is longer than the style. Thrum type heterostyly flowers experience poor set setting as the stigma remains hidden inside the corolla tube. This form ofstyle is found in almond

Top grafting: A method of grafting performed on top of the shoot.

Top working: A grafting procedure employed primarily to change the inferior unwanted cultivar of the established plant either by grafting (top grafting) or by budding (top budding).

Topiary: Developing a plant or training a plant into different forms or to shapes like animals, birds, seats etc. The shrubs which are amenable for bending, frequent trimming and training are suitable for developing topiary. Eg. Cupressus, Casuarina, Bougainvellesetc

Topophysis: The phenomenon in which different parts of the plant show phases variation and meristem perpetuate these different phases in their vegetative off-springs.

or

The phenomenon that occur when scion, budding and rooted cuttings maintain for sometime the branch like growth habit (plagiotropic growth) they had as shoots on the ortet.

Totipotency: The ability of a somatic cell to reproduce the entire plant somatically.

Training: It is a new practice in which tree growth is directed into desired shape and form.

Transgenic: A term applied to organisms that have been altered by introducing DNA molecules into them.

Tree spices: Clove, Nutmeg, Cinnamom, Allspice, Cassia Tamarind,Bay leaf, Curry leaf.

Tropical fruit: The fruit plants grown in region where no distinct summer and winter climate. This region receives high rainfall and has high humidity. Eg. mango, banana, papaya, pineapple, cashew

Truck garden: This refers to producing some special crops in relatively larger qualities for distant market. Cultivation is more extensive. The commodities are usually marketed throughmiddle men.

Tuber: It is a swollen, modified stem structure that functions as an underground storage organ. The eyes of the tuber present externally represent the nodes which are arranged spirillayeg. Potato, cassava.

Under stock: This term is used sometimes to describe rootstock especially in a double worked plants

Variety: An identifiable strain within a species, usually referring to a strain that arises in nature as opposed to a cultivar, which is specifically bred for particular properties, sometimes used synonymously with cultivar.

Vegetable: A product of herbaceous annuals which are usually consumed after cooking.

Vegetable forcing: It refers to the production of vegetables out of season in controlled atmosphere which may be a glass house, hot bed, hydroponic or an aeroponic system. This has more practical utility in countries with severe winter where tropical vegetable like tomato and cucumber are impossible to be cultivated in open. These vegetables are grown with condusive environment. Special varieties suiting protected cultivation have been developed in western countries. In India, such systems have little utilizes unless they are extensively produced for export which require adequate care on growing and post harvest operations

Veneer grafting: Method of grafting in mango, cashew nut where rot stock give slanting downward cut corresponding upward cut made on scion and unite together.

Vermiculite: Micaceous mineral and chemically hydrated magnesium aluminum silicate, very light in weight, insoluble in water and is able to absorb large quantities of water.

Virus indexing: Detection of virus infection of a plant by grafting or budding it on to an indicator plant known to be highly susceptible to that virus infection.

Warm season vegetable: They grows at a higher temperature range of 25 C to 35 C and are generally crops whose fruits are edible like solanaceous and cucurbitaceous vegetables except amaranthus and tapioca and sweet potato where other portions are consumed.

Wedge grafting: Synonymous to cleft grafting.

Weedicide: are the chemicals which are sprayed over the fields to get rid of weeds. Eg. 2,4-D acts as a dicot weedicide.

Whip and tongue grafting: A grafting method similar to splice grafting but another cut is made on both stock and scion beginning at the point about one third of the distance from the tip and about one half the length of the forst cut followed by the insertion of the scion in to the stock leading to maximum cambial contact.

Whip grafting: Synonymous to splice grafting where made a slanting cut on both stock and scion of similar age, diameter and then cambium of both plant parts unite together tightly with polythene strip.